Trail of the Wolf

R. D. Lawrence

TRAIL OF THE WOLF

KEY PORTER BOOKS

Canadian Cataloguing in Publication Data

Lawrence, R. D. (Ronald Douglas), 1921-
Trail of the Wolf

ISBN: 1-55263-149-4

1. Wolves. 2. Wolves – Pictorial works. I. Title.

Ql737.C22L39 599.74'442 C92-095687-4

The publisher gratefully acknowledges the support of the Canada Council for the Arts and the Ontario Arts Council for its publishing program.

Canadä

We acknowledge the financial support of the Government of Canada through the Book Publishing Industry Development Program (BPIDP) for our publishing activities.

Key Porter Books Limited
70 The Esplanade
Toronto, Ontario
Canada M5E 1R2

www.keyporter.com

Design: Scott Richardson
Cover photograph: Erwin and Peggy Bauer
Maps and Illustrations on pages 19, 27, 46, 47 and 149: Dorothy Siemens
Photographs: pages 1 and 3, © Tom and Pat Leeson; page 5, © John and Ann Mahan
Details: page 6, © W. Perry Conway; page 16, © Tom and Pat Leeson; page 42, © Wm. Munoz; page 70, © Thomas Kitchin; page 80, © E. A. James/NHPA; page 96, © Bob Gurr Photo; page 120, National Archives of Canada; page 146, © C. Allan Morgan

Printed and bound in Italy

00 01 02 03 6 5 4 3

CONTENTS

Introduction

SURROUNDED BY WOLVES

Early one morning in February, 1955, I had my first close encounter with wolves. It was an awe-inspiring experience, especially since I had taken up my homestead in northern Ontario only two months earlier. I had recently arrived in Canada from England with my head full of European mythology, all of which painted the wolf with the broad strokes of hatred and fear. Wolves were savage, cruel creatures that invariably attacked humans, according to most of the myths; others claimed that wolves killed more prey than they could eat; yet others swore that wolves had the power to turn humans into werewolves. Combined, these myths presented the wolf as the devil incarnate.

As a biologist, I did not believe all of those ridiculous stories; but, on that day, very soon after I entered the trail I had made a week earlier through the boreal forest, I started to think about wolves. Although I had travelled unconcerned along that same route about a dozen times, I became apprehensive when I found myself embraced by the dense stands of black spruce that reduced visibility to a few yards on both sides of the trail. Inexplicably, and for the first time, I felt totally alone. Unarmed except for an ax and a bow saw, and far from my nearest neighbor, I wondered if it was safe to be

Deep snow makes travel difficult for wolves, so they often use frozen snow-covered waterways and established trails to move from one area to another.
© J.D. Taylor

snowshoeing alone in wolf country without a rifle for protection.

The temperature was –20°F (–30°C) and knee-deep snow covered the ground. Soon after I entered the trail that led to the site where I had been logging spruce trees to sell to a pulp mill, I started to find the usual assortment of animal tracks. The first snow-prints had been left by the cloven hoofs of moose; then, further in, where the path narrowed, I paused to examine the tiny pad marks of red squirrels and mice and, now and then, the three-toed tracks of ruffed grouse.

After I had walked about a mile (1.5 km), I arrived at the logging site of one of my neighbors and stopped to chat for a few minutes. I noticed he had brought a rifle with him. It was dangling by its sling from the stub of a tree branch. My neighbor saw that I was looking at the gun, and he explained that he always carried it when his meat supply was low. "For moose," he said. "And if I happen to see a wolf, I shoot it. The pelt's worth a few dollars."

Moments later I resumed my journey. The sun was just showing itself above the trees and the clear blue skies promised a cold but fine day. Because I still had about a mile and a half (2.5 km) to travel in order to reach my own workplace, I began to hurry, mushing along without paying much more attention to tracks until I reached a sharp bend in the trail and encountered the unmistakable spoor of wolves. The wild hunters had crossed the trail quite recently, I thought, and there had been a number of them, perhaps six or seven, judging by the many tracks and by the slightly different size of each set. I forced myself to keep going, trying not to think about wolves.

In due course, I reached my logging site. I removed my snowshoes and slipped off the small haversack in which I carried my lunch sandwiches and a thermos of coffee. After setting these articles on

In a storm, when the wind moans through the trees or whistles through boulders, wolves are often stimulated to howl.
© Peter McLeod/ First Light

the pile of logs that I had cut and stacked during the past five days, I selected a spruce tree for cutting. Using the ax first, I chopped a notch on the fall side of the tree. Then, kneeling in the snow, bow saw in hand, I was about to start sawing when I heard branches snapping from somewhere nearby. Seconds later came two howls; they were short calls, each punctuated by a deep bark. Wolves!

By the time I stood up, dropped the saw, and picked up the ax, I realized that I was surrounded by wolves. They were circling me, while howl-barking, an alarm call that starts with a relatively high-pitched bark which almost immediately turns into a short howl. Nevertheless, they were keeping their distance. With my heart pounding, and the hairs standing up on the nape of my neck, I started backing towards the logs, a pile some five feet (1.5 m) high.

But as soon as I began to move, the howls intensified; they had been deep toned, now they rose to a high pitch. And there were more of them. I turned to face the logs and ran clumsily towards the pile. Once there, as I was scrambling up on the stack, I grabbed a short, thick pole that I had used as a measure. Scared and unsure of what I should do, I stumbled to my feet, clutching the ax in my right hand and the make-shift club in the left. By now I had caught glimpses of several wolves — large, gray-brown creatures that coursed through the snow swiftly and easily. I was quite sure that I was going to be attacked and almost certainly killed.

The wolves continued to race around me, although they remained mostly among the trees, rarely venturing into my small clearing. Now and then I would see one or two of them about twenty feet (6 m) away as they paused for a few seconds, stared at me, then darted back into the trees to continue their unnerving cacophony. I cannot tell how long I stood on the pile of logs clutching my puny weapons; it may have been ten minutes, or it may have been half an hour. What I do know is that it was the most fearful experience of my life. In the end, I could stand the stress no longer and decided that if I was going to be attacked, I would initiate the encounter by seeking to leave the area.

Screwing up my courage, I jumped down from the logs and hesitated, wondering if I should strap on the snowshoes. After a moment's thought, I decided against it, for I would have had to drop the ax and club and stoop over to strap on the webs, putting myself at a total disadvantage. Now, determined to get away, I waded through the snow to the trail entrance, moving slowly and trying to appear calm and purposeful. To my surprise, I reached the opening without

A pack socializes before the hunt.
© Wm. Munoz

incident and only then did I realize that as soon as I had started to walk towards the trail, the wolves had stopped howl-barking.

Had the pack decided to let me go?

Some twenty or thirty paces along, I stopped suddenly when I became aware that the forest was quiet, that the only sounds now reaching me were the susurrus made by the wind as it stroked the tree tops, the chirping of chickadees, and the distant cawing of ravens. Of the wolves there was no sign.

I felt somewhat reassured and plodded along, intent on reaching my neighbor's logging site. I planned to borrow his rifle and return to my site to continue my work. Eventually, I reached my destination. My neighbor had heard the wolves but, of course, had not realized that they had surrounded me. In any event, he lent me his rifle and a dozen cartridges, and I began to retrace my journey.

About two hours after I had been chased away by the wolf pack, I re-entered my small clearing holding the rifle at the ready. Despite the weapon, I felt apprehensive, particularly so when I saw that many wolf paws had trampled the snow within my working area. The wolves had obviously investigated the log pile and had seemingly paid special attention to the tree that I had been about to cut down when they disturbed me. One or more pack members had urinated against the notch that I had chopped in the tree, as well as against the log pile and on the bow saw, which I had left lying on the snow.

Now my deep interest in biology gradually took over. I was no longer afraid (undoubtedly because of my neighbor's rifle) as I began to sort out the tracks. At first I was somewhat confused by the haphazard trampling, but presently I decided that the wolves had left as a group travelling in an easterly direction. I followed their tracks for

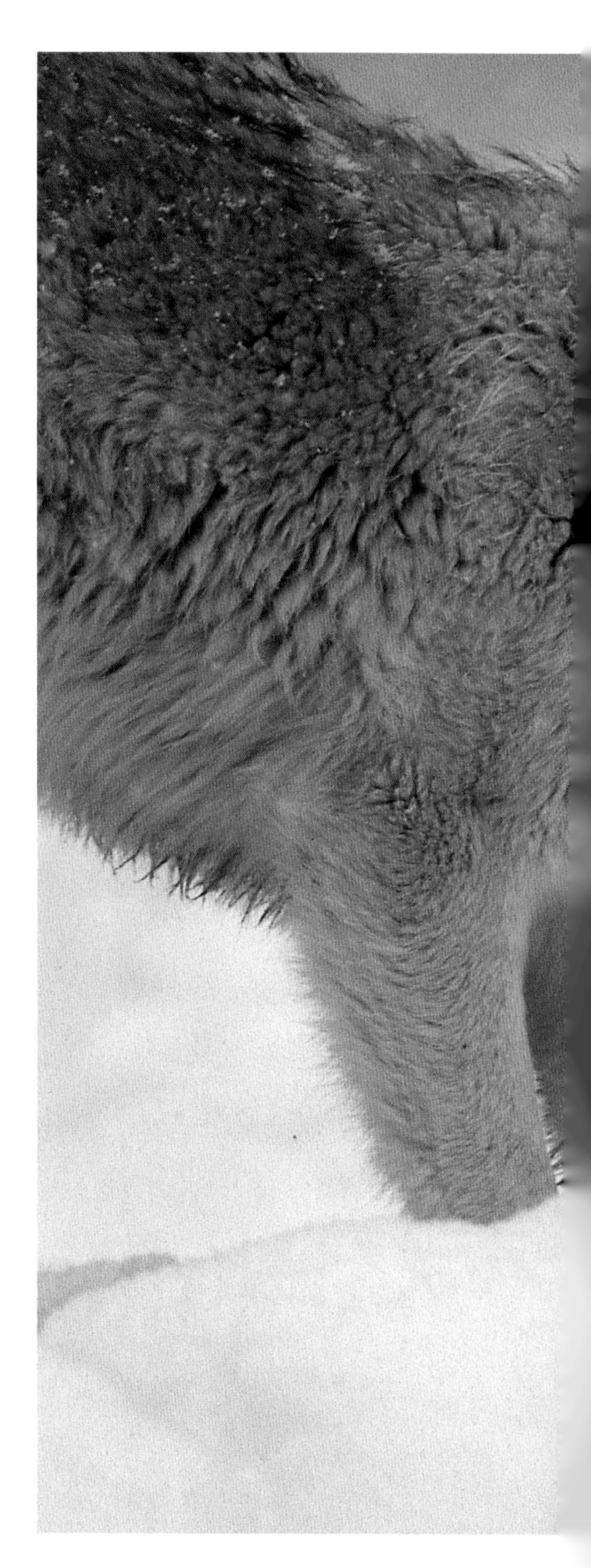

An argument over the remains of prey. An adult wolf can consume about twenty pounds (9 kg) of meat in one feeding, although the average daily amount is much less.
© W. Perry Conway

a short distance until, some ten feet (3 m) ahead, I was startled to see numerous bloody marks on the snow. I walked on slowly and cautiously and, a moment later, encountered what was left of a white-tailed deer.

The snow around the deer's remains was greatly disturbed. Mounds of reddish-white hair were scattered about a rough circle. In one place I found the deer's head. Some skin and fur still adhered to it, but the brain pan had been torn open and its contents removed. One eye was missing; the other had been punctured. I counted nine large punctures on the bone, which had been made by wolf fangs.

I now realized that I had arrived at the logging site either immediately after the wolves had killed the deer or when the pack had already started to feed on it. It came to me that the pack had displayed great forbearance, for to blunder upon a group of carnivores just as they were eating, or about to eat, was asking for trouble. Yet my arrival did not provoke an attack. Instead, they had simply frightened me away.

My fear of wolves evaporated that day. Well, almost — for mythology has a way of burrowing deep into the psyche and, once entrenched there, it departs with great reluctance.

When I returned the rifle to my neighbor later that day, I told him about finding the deer's remains and confessed that I had been truly afraid during the confrontation. He smiled, nodded, and told me that after having spent thirty-five years in wolf country, he had never once been so much as threatened by wolves.

"You know, you're lucky to have seen them. Wolves keep well clear of humans, as a rule. I've only seen singles now and then, and every one of them sure disappeared quickly when they noticed me."

My encounter with the pack impressed me greatly and fired my interest in wolves. How, I wondered as I snowshoed home, could I study them?

Nearly a month later, while returning home through the forest at sunset, I felt I was being followed. Once again I became apprehensive. I was not actually afraid this time, but I suppose I was nervous knowing that some creature I could not see was tracking me.

On that occasion my only weapon was a large hunting knife and, as before, I was entirely alone and far from my own kind. But, remembering my unwarranted fear of the wolf pack, I calmed down and allowed common sense to take over. As a biologist, I was very curious. I wanted to see whatever it was that was following me.

I stopped and turned around quickly. Nothing was behind me. Had I made a mistake? I resumed walking and sensed almost immediately that I was being tracked. I took a few more paces then swung around, this time without stopping. I was just in time to catch a fleeting glimpse of a wolf as it bounded away from the trail. I became fascinated. Why was the animal following me? It was obvious that it was not disposed to attack me. Had that been its intention, it would have done so already. Then I remembered that one of the two meat sandwiches that I had taken with me for lunch had remained uneaten and was in the small backpack that I was carrying. Was the wolf able to smell the cooked meat? Or was it, perhaps, interested in the odor of my snowshoe webbing?

These questions nagged at me as I resumed my journey and again felt the presence of my follower. Soon afterward, when I was almost at the entrance to my home clearing, I removed the sandwich from my pack and left it on the trail. Then I continued onwards,

The wolf has the greatest natural range of any living land mammal other than man.
© Tom and Pat Leeson

reached my clearing, and started across it. But I stopped after about fifty yards (46 m) and looked back. A wolf was sitting at the edge of the forest, looking at me. It was a large animal with a dark gray coat which shone as though polished. In the slanting rays of sunlight, the animal looked almost silver, a tone that contrasted with the black, saddle-shaped patch that decorated its shoulders. The wolf raised its head, revealing a white star high up on its chest. We watched each other for some moments, but when I decided to approach the animal, it rose to all fours, stared at me for a moment longer, turned, and trotted into the forest. I walked back to see if the wolf had eaten the sandwich. It had.

Already stimulated by my experience with the pack, this second close encounter helped me decide then and there that, despite my commitment to the general study of nature, I would now also concentrate on wolves, difficult though such a task might be. Of a certainty, I knew at that moment that I would never again be afraid of wolves.

That evening, dwelling upon my two lupine encounters, I realized that I had already learned a couple of facts about wolves. In the first place, the wolves in the pack had demonstrated a keen intelligence; in the second place, the loner that had followed me, I was sure, had done so because he could smell something about me that attracted his interest. If I was right, the animal possessed an extremely keen sense of smell.

Today I know that I was right on both those counts, but it has taken me thirty-seven years of studying wolves in the field and in captivity to realize just how highly intelligent these magnificent mammals are and how acute their sense of smell and their other incredible faculties are.

Chapter One

WHAT IS A WOLF?

One thing I quickly learned when embarking on my studies of the wolf was that there was a good deal of confusion regarding the ancestry, number of subspecies and common names of my favorite animal.

Paleontological evidence strongly suggests that the wolf's ancestor was a relatively small carnivorous mammal that evolved during the upper Miocene epoch, some fifteen million years ago. It is believed that by means of a long series of mutations, this primitive creature, named *Tomarctus*, eventually gave rise to wolves, foxes, and a number of their near relatives, including the domestic dog. In fact, the family Canidae, to which wolves belong, contains sixteen genera (or kinds) of carnivores (meat-eaters). These in turn have been separated into thirty-six species, which are widely distributed across most of the terrestrial regions of the world.

The living members of the Canidae, or dog family, have been further divided into three subfamilies based mainly on their dentition: the number of teeth, their shape, and their size. The first Canidae subfamily, and the largest, is the Caninae. It contains the wolves, the dogs, the coyotes, the jackals, the foxes, the raccoon dog, the maned wolf, the bush dog, the small-eared dog, and the dhole.

The two species of wolf — the gray wolf (*Canis lupus*) and the

A gray wolf leaps over a fallen branch in early summer.
© Erwin and Peggy Bauer

red wolf (*Canis rufus*) — contain the largest members of the Caninae. They, in turn, have been divided into a number of subspecies by biologists, some of whom have been prepared to name a subspecies after finding one bone that is different from the same bone found in other wolves. In this way, a number of subspecies claims were made which have since been refuted. At one time, for example, it was believed that twenty-five subspecies existed in North America. Four subspecies of the gray wolf were listed for the arctic islands, nine were said to live on the tundra and in Newfoundland (they have been extinct on that island since 1911), seven were said to live in the western mountains and along the Pacific coast, and two were noted for the eastern and central regions of North America. Three subspecies of the red wolf were said to inhabit the Mississippi valley, Texas, and Florida.

In more recent years, the scientific literature notes that nineteen subspecies exist, or existed, in Canada, the United States, and Mexico, and eight in the Old World. One of these subspecies, which inhabited the central regions of the United States and Canada, was *Canis lupus nubilus*, the so-called "buffalo wolf," which preyed on bison. It became extinct before the turn of the century, but there are some unsubstantiated claims maintaining that small numbers of these wolves have survived. Another subspecies, *Canis lupus lycaon*, the timber wolf, although mercilessly extirpated in most of the continental United States and southern Canada, continues to survive in the forested areas of the eastern and northeastern regions of its ancestral range.

Subspecies confusion would appear to hinge upon the fact that biologists, somewhat like paleontologists, are divided unevenly into

The evolution of the wolf.

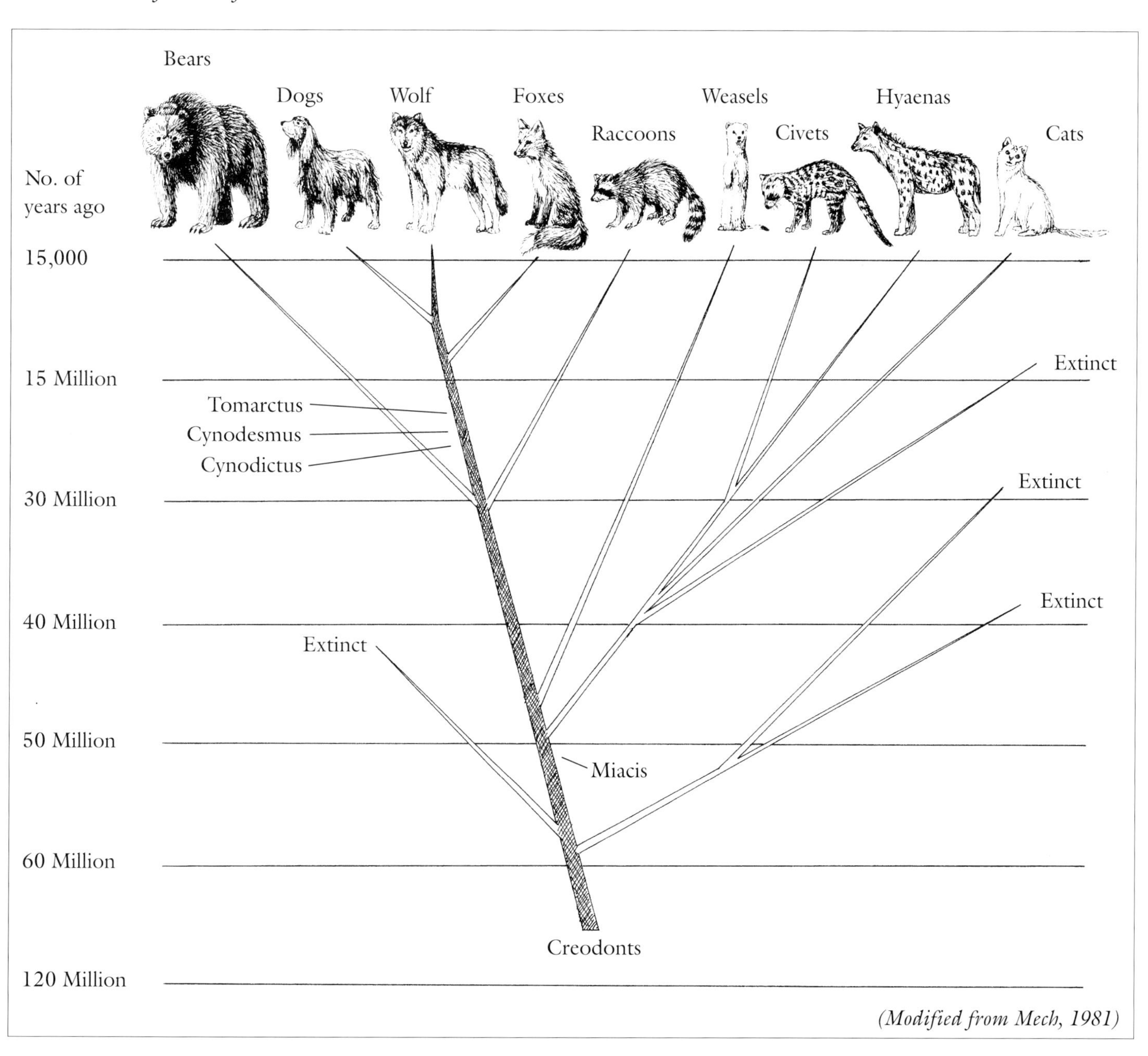

(Modified from Mech, 1981)

two groups — the "lumpers" and the "splitters." Splitters recognize a sometimes-large number of subspecies while lumpers seek to simplify the family tree of a genus by recognizing fewer subspecies. In any event, when it comes to the scientific names of subspecies, whether a wolf is named *Canis lupus tundrarum* (the tundra wolf of Alaska and northern Canada) or *Canis lupus columbianus* (the British Columbia wolf) or *Canis lupus lupus* (the Eurasian wolf), to paraphrase Shakespeare, a wolf by any other name is still a wolf.

A red wolf pup is caught by researchers of the U.S. Fish and Wildlife Service at a captive breeding station.
© Tom and Pat Leeson

Another source of confusion is the definition of a species. In his book *Principles of Systematic Zoology*, biologist Ernest Mayr noted that a species is a group of natural populations that are capable of mating and producing viable, fertile offspring. When applied to the wolf, such a definition can be puzzling because one cannot determine with any degree of certainty the identity of the original species. Did *Canis lupus* sp first emerge in the Old World or did it make its debut in North America? In all probability, the first true wolf emerged in Eurasia. Paleontologists have discovered evidence of several kinds of wolves living in Europe during the Villafranchian mammal age about two million years ago.

In the final analysis, it is difficult to say whether the original wolf species continues to survive, because inbreeding on a grand scale has been occurring for a considerable length of time. Today inbreeding between the various subspecies continues at a rate faster than ever before. The major cause is the persecution of the wolf by people throughout its range.

Mating between individuals of two different, but allied, species can also occur. This is called hybridization. One example is the deliberate crossing of a horse and a donkey to produce a mule, which is

A red wolf is fitted with a radio collar for research purposes.
© Denver A. Bryan

infertile and cannot breed. There is now speculation that the red wolf may be a hybrid that is capable of breeding.

During the last decades of the nineteenth century and the early decades of this century, the red wolf, like its northern counterpart, was considered to be a serious threat to livestock and was hunted, poisoned, and trapped to near extinction. The few survivors were eventually placed on the endangered species list in the United States and were given federal protection. Spearheaded by the United States Fish and Wildlife Service, a movement was begun to reintroduce the animal into at least part of its original range. This, as might have been expected, received a great deal of criticism from wolf haters, but the government officials continued to make their plans.

Bearing in mind the historic range of the red wolf, Fish and Wildlife Service biologists became interested in 120,000 acres (48 500 ha) of swamp and marsh in coastal North Carolina that had been donated to the federal government by the Prudential Life Insurance Company in 1984. They decided that this donated land, which had been proclaimed the Alligator River National Wildlife Refuge, would make an ideal homeland for red wolves. And so, the first four pairs, which had been raised in a captive propagation program, were released in the refuge in 1987.

In June, 1992, two scientists, R.K. Wayne and S.M. Jenks, published their findings on the genetic composition of the red wolf in *Nature*, the British scientific journal. Their examination of DNA obtained from captive red wolves and the skins of museum specimens captured at the turn of the century indicated that the animal could be a hybrid — part gray wolf and part coyote. This may or may not be the case and, indeed, the reliability of genetic testing has been

The maned wolf of South America is not, in fact, a wolf, but a separate species belonging to the Canidae family.
© G.D.T. Silvesbris/ NHPA

questioned by some biologists. Nevertheless, the implications are important; if the red wolf is not a separate species, it may no longer be protected by the *Endangered Species Act*, and, in fact, after the publication of the study, ranchers in Tennessee and North Carolina petitioned the government to have it removed from the list. Their petition was denied, and, for the time being, although more people are questioning the continuation of the Red Wolf Recovery Plan, no steps have been taken to end the program.

There has been some speculation that a red wolf is a cross between a gray wolf and a coyote.
© J.D. Taylor

To add further to the species/subspecies confusion, it has been found that red wolves and coyotes have mated, at least to some extent. In recent years, because the numbers of red wolves have been drastically reduced, the coyote has expanded into those regions vacated by the other animal. Mating between individuals of the two species seems to have occurred in areas where their ranges have overlapped. One must now wonder whether a new subspecies has emerged. If so, is it a subspecies of red wolf or a subspecies of coyote?

There is also evidence to show that in northeastern parts of the United States and southcentral Canada, wolves and coyotes have mated to produce large hybrid carnivores. However, it appears that such crosses have been few, probably because wolves in those regions have been mostly exterminated by trapping, poisoning, and hunting.

At this stage one may well ask why it is so important to be constantly seeking out new subspecies? Many lumpers view the splitters' zeal for taxonomy (scientific classification) as an unnecessary and rather esoteric aspect of biology, while the splitters see "lumping" as an easy way out of a scientific dilemma. Nevertheless, taxonomy is useful if not carried to extremes, because it can delineate the pathways of evolution and, of course, the understanding of ecology. For

This coyote–red-wolf hybrid resembles the coyote more than it does the wolf. © Erwin and Peggy Bauer

this reason, the importance of scientific names cannot be overemphasized. They immediately and correctly identify an animal, whereas the common names can sometimes be confusing.

In the case of the wolf, the common names of animals that are only distantly related to it can be misleading. This is certainly true of the maned wolf, which is found in Argentina, Bolivia, Brazil, and Uruguay. This animal is not a wolf. Rather, it looks like what might appear if one crossed a coyote with a red fox. It has long, black legs; a golden coat (except for a black mantle along its neck); a light-colored, relatively short, bushy tail; large ears; a long, sharp muzzle; and small feet. It weighs between 40 and 50 pounds (18 and 23 kg).

A large coyote in prime condition. Coyotes have expanded their range into former wolf territory.
© C. Allan Morgan

Confusion also arises when the coyote is referred to as a "brush wolf." Although it is a relative of the wolf, it is definitely not a true wolf. Use of the term "brush wolf" has created problems of identification in large areas of Canada and the United States, because people who use the term mistakenly believe that the "brush wolf" and the coyote are two different kinds of carnivores. Farmers who have lost calves, sheep, or chickens to coyotes are then quoted in the local media calling for the elimination of "wolves." In more recent years, coyotes and feral domestic dogs have been mating and have produced a hybrid animal that is larger than the coyote. Depredations made on farm animals by these "coydogs" are also said to be the work of "brush wolves." Because of the incorrect labelling of other animals as wolves, wolves frequently end up with bad press they do not deserve.

Physical Characteristics

The wolf's scientific name, *Canis lupus*, was originally coined in 1758 by noted Swedish scientist Carl von Linne, who is today better known as Linnaeus. Translated, *Canis lupus* means "dog wolf," which is somewhat ironic in that it was the wolf that gave rise to most of the domestic breeds of dogs rather than the other way around.

Opposite*: The howl of the coyote is higher pitched and more repetitive than the long-drawn-out, deeper call of the wolf.*
© Erwin and Peggy Bauer

Wolves and dogs share a number of traits, such as the length of the gestation period, molting in spring and growing a winter coat, and the order of appearance of the first teeth. In other respects, however, the wolf is quite different from the dog. In appearance, it has ears that are relatively short, broad at the base, and less pointed at the tips than those of most dogs. Its head is large, wide, and

Like wolves, the Cape hunting dogs of South Africa are highly social canids.

heavy, and the forepart of the skull curves downward to blend into a broad but tapering muzzle that ends with a black nose, which may be as wide as 1½ inches (4 cm).

Although the bones of the lower jaw are not as wide as those of the upper, both jaws are thick and strong and have enormous biting and holding power. Wolves usually have twenty-one teeth in each jaw (six incisors, two canines, eight pre-molars and five molars), but, as is not uncommon in many mammals, including humans, some wolves may have a few more or a few less teeth than others, the difference being contingent upon the actual length of the jaws, which can vary from wolf to wolf.

The legs of wolves are longer than those of most dogs, and the feet are larger, the front paws being longer and wider than the back feet. On average, the front paws of a large male wolf at rest measure 4½ inches long by 3¾ inches wide (11.5 cm by 9.5 cm). The back paws at rest measure 3¾ inches long by 3¼ inches wide (9.5 cm by 8 cm). When walking, or when running in sand, mud, or snow, wolves leave tracks that are considerably larger than when their feet are at rest. This occurs because on contact with the ground the foot spreads, elongating the toes and widening the pads. Thus, large wolves running or trotting on soft terrain may leave front tracks that are 5½ inches long by 4¼ inches wide (14 cm by 11 cm); the spoor left by the back feet would be about 20 percent smaller.

Wolves have five toes on each front foot and four on each hind foot. In fact, in all canids except the African hunting dog, which has four digits on each foot, there are five toes on the front foot and four on the hind. Unlike humans, who set their whole foot on the ground when they walk, or cloven-hoofed animals, which walk on their nails,

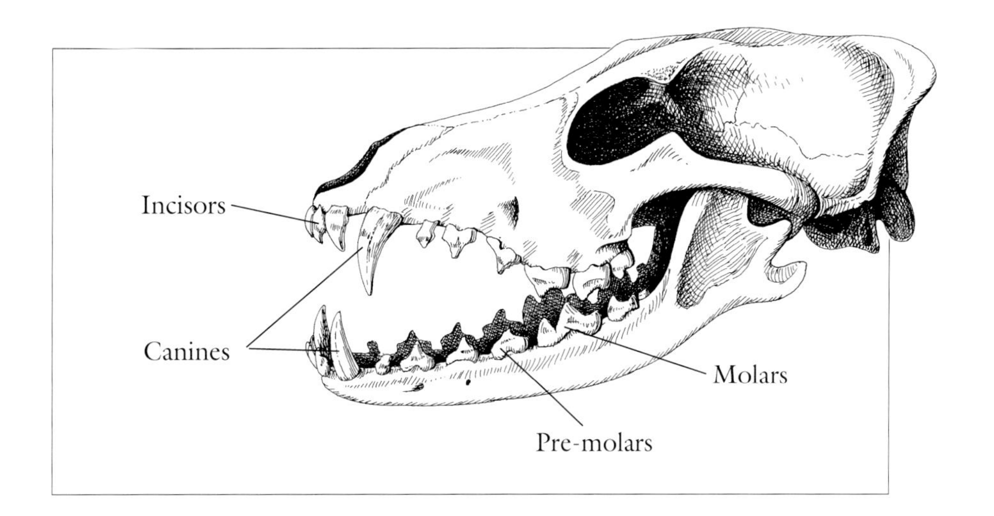

Wolf skull and teeth.

Wolf tracks. The bottom impression was made by a back paw, the top one by a front paw.
David Mech

wolves and other canids walk on their toes. Wolf tracks, however, show the impressions of only four toes on each front foot. This is because the "thumbs," which are located above the foot pads but below what in humans would be the wrists and the ankles, are vestigial and are not used for locomotion.

The nails on these vestigial digits do not get worn down from use and become curved and sharp-pointed. Dog owners often have them removed from their pets because of their sharpness. In wolves, these claws may be used when a wolf attacks an animal, but barely manages to hold it with its teeth. In this situation, if a wolf can manage to curl one front paw around the prey's leg, or if it can grab part of the animal's body with both front feet, one on each side, the so-called dewclaw digs in, helping to secure the hold and to bring down the prey.

Those big, springy feet help wolves move quickly. Depending on the terrain and the urgency of their quests, wolves are able to run at a top average speed of twenty-four miles (39 km) per hour for a relatively short time, perhaps three or four minutes, without becoming exhausted. In one instance, an unconfirmed report noted that a wolf running along a backwoods road in Minnesota had been clocked by the pursuing vehicle at a speed of thirty-five to forty miles (55 to 65 km) an hour for a quarter mile (0.5 km). As a rule, however, wolves travel at five to six miles (8 to 10 km) an hour when moving from one place to another or when patiently trailing prey — a rate that they can keep up for hours at a stretch. But whether running or loping, the movement of wolves is always smooth, their stride elastic and seemingly effortless as they cover the ground.

Depending on their range and subspecies, as well as on the

prey they hunt, the total body length of typical adult male wolves in North America and Eurasia, not counting the tails, varies from about four feet to almost six feet (1.6 to 1.8 m). Females are about five inches (13 cm) shorter. Tail length varies from fourteen inches to twenty inches (36 to 50 cm). Males are usually, but not always, about 15 to 20 percent larger and heavier than females. The average weight of adult male northern wolves is about eighty pounds (36 kg). However, an exceptional 175 pounds (79.5 kg) was recorded in Alaska for one wolf that had been shot and weighed when its stomach was full of meat. In general, northern wolves are considerably larger than their more southerly counterparts. The largest North American wolves live in western Canada and Alaska; the smallest in southeastern Canada and Mexico.

The differences in size between northern and southern wolves occur for two basic reasons, although a third factor may simply be genetic inheritance.

The first important benefit that a larger size gives northern wolves is heat conservation. The greater the volume of an object, the more heat it can conserve. For example, a cup of hot water will cool faster than a bathtub full of water that starts at the same temperature. Indeed, for this reason most northern-dwelling mammals, including mice and voles, are larger than their counterparts in more southerly areas.

The second benefit of a larger size is the strength, weight, and speed it provides. In the case of wolves, these qualities are needed to chase, capture, and kill large prey, such as moose, elk, and caribou, the bulls of which may outweigh a wolf as much as ten to one. The southern wolves in Eurasia, Arabia, India, and other warmer regions, on the other hand, tend to hunt smaller prey or, in some cases, have

The northern gray wolf (right) *differs from the southern red wolf* (above) *in several ways: it has a wider body and head, and longer fur.*
© Tom and Pat Leeson (*right*); © Mike Biggs (*above*)

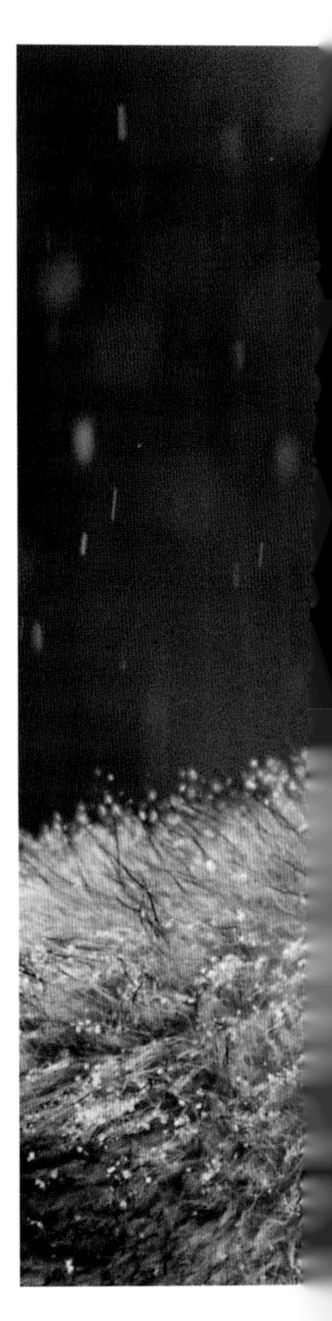

A wolf sleeps during a snowstorm. The accumulating snow will insulate it from the cold.
© Tom and Pat Leeson

been forced to depend on garbage and domestic animals.

The coats of northern wolves and southern wolves that live in hot climates are also different. The southern subspecies have thinner coats, which help them to keep cool. Conversely, at the approach of autumn, northern wolves grow heavy coats that are composed of shiny, long, moisture- and dirt-shedding guard hairs and thick, wool-like underfur. Like sheep wool, the underfur contains an oily substance akin to lanolin, which makes it impermeable. This type of coat keeps its owner warm, even in the Subarctic and Arctic, where north of the Arctic Circle the sun does not rise above the horizon throughout the winter.

The northern wolf's feet are perhaps the most vulnerable to the cold of all its body parts, but even these are protected by fur that grows between the toes and around the pads. The pads are also equipped with many heat-producing blood vessels. When the wolf is trotting or running, the accelerated circulation of the blood keeps its feet warm. When resting during the most frigid times, wolves curl up in a tight ball and use their bushy tails to cover their heads and ears. On the open tundra, because the snow blows about almost constantly, a sleeping wolf will soon be covered by an insulating mantle of snow.

A wolf's guard hairs, apart from shedding moisture, keep its coat relatively free of burrs and dirt because of their hard, smooth, slippery composition. If, for instance, a wolf rolls in carrion or mud and rises caked with clots and clods adhering to its coat, it shakes itself vigorously, ridding itself of the bigger lumps of dirt. Then it waits until the remaining material is dry, when it again shakes itself, causing most, if not all, of the particles to fly off.

The body odor of wolves, unlike that of most dogs, can only be

This wolf has a typical spring molt appearance. It is starting to shed its long guard hairs and woolly underfur.
© Thomas Kitchin

detected by the human nose if it is pressed almost directly into the fur. Prey species, whose olfactory system is more acute than that of humans, can probably pick up the scent of wolves over relatively long distances. This may explain why wolves invariably try to smell like something other than themselves by deliberately rolling in carrion — a useful habit for animals that hunt prey whose sense of smell is almost as acute as theirs. It is this ancestral trait that continues to linger in the genes of most dogs, much to the disgust of their owners.

Although the coats of southern wolves in North America and Eurasia are lighter than those of their northern counterparts, they do grow winter coats composed of guard hairs and underfur, but the woolly fur is finer and may hardly grow at all in hot regions, such as the Sinai or Arabian peninsulas.

In the spring, which, depending on the region, may vary from March to May, wolves start to shed both their guard hairs and their underfur. In northern regions, molting produces enormous quantities of hair. As the old coat is being shed, the new coat begins to grow. This metabolic event uses protein and causes the wolves to become rather gaunt during their time of molt. In the end, they will have entirely replaced their coats, although the new underfur will not become really thick and long until late autumn, which begins in early September in the extreme north and progressively later at lower latitudes.

The arrival of the warmer season, and especially the hot summer, can be trying for northern wolves. They usually hunt at dusk or dawn. During the day, in regions where forests are available, they sleep or relax in the shade of the trees until the cool of evening, even if they have not been successful on the day's hunt. On the tundras of North America and Eurasia, however, wolves seek shelter in areas of heavy shrub, or they may lie in the shade of hillocks, especially the permafrost upthrusts known as pingoes. Pingoes are ice-cored, dome-shaped elevations which stand from about six feet (2 m) to as high as 150 feet (46 m). They rarely melt, so they act as natural refrigerators. Indeed, the native Inuit dug into pingoes and turned them into impromptu ice-houses in which to keep their meat.

Regardless of where wolves may lie down for any length of time, their bedding places can usually be recognized by the depressions left by their bodies in the shrubs or grasses, and by shed fur, which is especially evident during the molt, when great handfuls blow about the area or remain in loose-tangled clumps on the crushed plants. In winter, bedding sites appear as round depressions in the snow. Ice, formed when the animal's body heat melts the snow, will be found in the depressions. Wolves do not use the same bedding site again, because the layer of ice does not allow them to settle their hips comfortably.

The color of a wolf's coat is extremely variable, particularly in northern latitudes. Some wolves are snowy white; others are black except for a white blaze on the chest; others may be gray; and yet others have coats that contain mixtures of fawn, cream, gray, white, brown, and black. But all wolf coats, regardless of color, are decorated by what is referred to as a "saddle," a mantle over their front quarters shaped somewhat like a horse saddle. It is composed of extra-long guard hairs, which differ sufficiently in color from the rest of the coat to be easily identified. The function of the saddle, if it has any beyond pure decoration, is unknown; however, because it is individually distinctive, it may provide visual recognition among pack members. When a wolf is aroused and raises its

Although many wolves are gray, they can also be reddish, white, black, and most shades in between. © Erwin and Peggy Bauer (*right*); © Thomas Kitchin (*above*)

hackles, the long hairs on the saddle become immediately visible.

In addition, some wolves — and perhaps all of them — change the tone and even the color of their coats during each annual molt, especially as they begin to age. For instance, when a yearling captive wolf grew his adult coat, it emerged black except for the white star on his chest that many black wolves display. During the following year's molt, however, his coat was liberally "salted" with white hairs and, from the eyebrow area down the bridge of his nose, he showed a definite "widow's peak." In like vein, a companion female whose coat was a mixture of white, fawn, russet, and black, became much lighter. After that first molt, the changes occurred regularly every season and now, at eight years of age, the male is light gray and the female is almost entirely white.

The same holds true of the captive pack kept by Jim Wuepper of Negaunee, Michigan. Shawano, the alpha male of the pack, became almost entirely white by the age of seven, and the black alpha female became light gray at the age of six. Why such color changes occur is not known.

Threats to Survival

At this stage, it is useful to emphasize that individual differences are common among mammals of all species, including humans. Such differences involve the physiology as well as the behavior of individuals. It is for this reason that I dislike generalizing about the biology and behavior of wolves, or, for that matter, of any other species of animal, although I do resort to some generalizations for the sake of brevity and to give an overall concept of the animal, its habitat, and its prey.

A Mexican wolf photographed in captivity. There are thought to be only about ten Mexican wolves remaining in the wild.
© Mike Biggs

One must, for example, generalize when discussing the lifespan of wolves, because accurate figures are only available for captive individuals. Such animals, if kept in good condition and in roomy enclosures, live between eleven and twelve years on average but have been known to live as long as seventeen years. In the wild, however, the lifespans of wolves are shorter. They must work hard in order to keep themselves fed, and they often hunt through very rough country, where they risk injury from tree snags, rocks, and unexpected rock falls, or risk drowning in fast-flowing waterways.

At times wolves are also mortally injured by the very animal that they are attempting to pull down, especially when hunting large prey. For example, eye-witnesses have reported on several occasions that moose with a wolf clinging to their sides or their shoulders have been able, by dashing through closely packed trees, to brush the animal off against a large trunk. Such a collision may kill the wolf outright, or it may inflict injuries so severe that the animal will eventually die.

Wild wolves face other dangers in addition to accidents. When they eat the internal organs of their prey, all of which carry one or more parasites, wild wolves ingest these organisms and become infested by them. Indeed, wolves and coyotes are prey to a large number of internal and external parasitic species during temporary infections, although not all at the same time of course. These include protozoans (unicellular organisms), trematodes (flatworms), nematodes (roundworms), cestodes (tapeworms), acanthocephalans (spiny-headed intestinal worms), mange mite, lice, ticks, and fleas (mostly on coyotes). Little is known, however, about the effect of most of these parasites on the wolf population. Nor is much known about distemper and other diseases affecting wolves.

Few wolves and coyotes fall victim to rabies in North America. In a paper entitled "Host-Parasite Relationships in the Wild Canidae of North America," written by Danny B. Pence and J.W. Custer, the scientists note:

> ... the enigma of wild species of the genus Canis and rabies virus continues to defy elucidation. Although rabies has been described as a 'potential killer of coyotes' (Gier et al. 1978), only occasional cases of rabies in coyotes are reported in most western states of the United States.... Of the 9,943 laboratory-confirmed rabies cases from the United States, Canada, and Mexico in 1977, there were only three cases involving timber wolves in Alaska and one case each involving coyotes from Canada and Mexico (Center for Disease Control 1978). In rabies surveillance programs reported by the Center for Disease Control, Atlanta, Georgia, 12 of 435 (3 percent) coyotes in the United States examined had rabies, while 6 wolves in Ontario and one wolf in the Northwest Territories were found positive for rabies (CDC 1978).
>
> Of the 8,598 rabies cases in Ontario, Canada, from 1961 to 1969, only 35 (0.41 percent) involved coyotes and/or wolves (Johnston and Beaureagard 1969).... The prevalence of rabies virus and its role in the mortality of feral canids is in need of clarification.

Wolves feed together on the carcass of a deer. When their prey is large, several pack members eat at the same time.
© Thomas Kitchin

Recently, according to a 1992 report in the *Journal of Wildlife Disease*, outbreaks of Lyme disease have occurred in gray wolves from Minnesota and Wisconsin. The paper notes that researchers tested 528 wolves between 1972 and 1989 and isolated the antibodies for the disease in 15 (3 percent) of the tests. Although a 3 percent positive result seems low, it does indicate that wolves are, or have been, exposed to the disease and are therefore considered to be susceptible to it. It is not known whether the disease is fatal to wolves, but there is no doubt that they are being affected by it to some degree. Lyme disease, which is spread by a deer tick, *Ixodis dammini*, also attacks humans and dogs, and was first discovered in the town of Lyme, Connecticut. It has not hitherto been known to infect wolves. However, the disease appears capable of spreading to wherever a tick-carrying animal may migrate. In this context, Lyme disease has infected people in England in recent times, presumably because animals imported into the country from North America were carrying the disease or the infected ticks.

Finally, some wolves suffer from a condition affecting the liver-shaped sebaceous glands that are present in the skin surrounding the rectums of wolves and dogs. The glands at times form benign tumors, which are prone to rupture. In some dogs, the tumors will also form along the midline of the abdomen, but in some wolves, they form along the back, usually on one or the other side of the spine. When a tumor ruptures in a well-cared-for dog, veterinary treatment will usually follow. In wild wolves, however, if one ruptures, flies are sure to find the openings and lay eggs in them, maggots develop, and the cyst becomes infected. Without treatment, septicemia develops, and the animal's chances of survival are poor.

Of course, not all of these problems beset wolves at the same time. Nevertheless, the fact that wolves manage to survive in their environment while plagued by so many different threats to their well-being speaks highly of their incredible stamina and resistance to disease.

Wolf Senses

When I am asked to name the most important of the wolf's five senses, I usually answer "smell." But then I explain that, although the olfactory ability of wolves may well be its most important survival trait, sight, hearing, touch, and taste are probably equally important, because, in combination, they keep the wolf informed of all events occurring in its immediate environment.

Nevertheless, wolves do have an incredibly acute sense of smell due in part to their specialized olfactory systems — their fine-tuned noses — and in part to the complex functions of the mammalian endocrine glands. These include the pituitary, which excretes a variety of hormones; the sex glands, which excrete testosterone in males and estradiol and other hormones in females; the adrenals, which excrete epinephrine (adrenalin); and the thyroid, which produces thyroxin. All these glands, and a number of others, secrete their hormones directly into the bloodstream.

Endocrine hormones are continuously excreted by a mammal through urine, feces, and the skin pores, especially the pores on the pads of the paws (and the hands and feet of humans), which have more scent gland surfaces than any other part of the body. These excretions, called pheromones, are composed of minute traces of hormones that are harnessed to sulfur molecules. This combination evaporates quickly and is capable of being detected by insects and mammals (excluding humans) over long distances.

Each individual has a pheromonal scent signature inherited from both parents in equal or unequal proportions. That is to say, an individual may inherit more genes from one parent than from the

A wolf rolls on carrion. This custom, also found in domestic dogs, is intended to camouflage a wolf's own scent, which could alert prey.
© John and Ann Mahan

other, or, conversely, may have an equal proportion from each parent. Wolves can detect the pheromones of their own kind, or of the other animals with whom they share their range, over a distance of as much as 1½ miles (2.5 km), and probably farther away than that if the breeze is favorable.

Some years ago, while I was studying wolves in the region of British Columbia's Spatsizi Plateau, the pack that I had been observing for five weeks picked up the scent of a moose, which was about a mile (1.5 km) away, at the far end of a relatively narrow valley. The wolves were on high ground but could not see the moose because of the heavy forest that lay between them and the ungulate, but I was sitting about 200 feet (60 m) higher up on a ledge above the pack. As a result, when first the lead male and then all the seven wolves that made up the pack turned their noses down the valley, I trained my glasses in the same direction and almost at once saw the moose, a large bull that was belly-deep in a small lake, eating aquatic vegetation. Immediately after locating the animal, I swung the glasses back onto the wolves and was just in time to see the male and female leaders change direction in order to investigate the scent that had obviously attracted their attention.

I watched the pack follow the scent in an almost straight line. On reaching their quarry, however, the wolves were unable to make the animal leave the water, succeeding only in causing the bull to adopt a threatening attitude. So, the wolves, being opportunists, soon turned away and disappeared in the distant forest.

In addition to the sensitivity of their noses, wolves have acute hearing. They have been known to respond to humans imitating wolf howls from as far away as three miles (5 km). Their ears are

continuously attuned to every sound in their environment, even, it seems, when they are asleep. Although the rest of the animal may be slumbering and fully relaxed, the ears appear to be on the alert and quick to arouse a sleeping wolf if a suspicious or interesting sound is detected.

Wolves have relatively poor frontal vision, perhaps being unable to recognize clearly the features of their own pack mates at a distance beyond 100 to 150 feet (30 to 45 m). Their myopia evidently stems from the absence of the *fovea centralis*, the tiny pit at the back, center of the retina which, in humans, primates, and some other animals provides the point of sharpest vision. Precisely how clearly a wolf sees when looking directly at an object is, of course, impossible to determine accurately; humans simply cannot see through the eyes of a wolf! But in the absence of the *foveas*, it seems evident that beyond a relatively short distance their vision must be somewhat blurred, rather like that of a photograph taken with a wide-open lens at a slow shutter speed, as opposed to an exposure taken with the smallest lens aperture at a fast speed.

Nevertheless, wolves can see shapes and, especially, movement over long distances, and their peripheral vision is extremely acute. They are able to detect even the slightest movements of very small animals, such as a mosquito, at a distance of more than ten feet (3 m) and the movement of large animals at considerable distances.

It has long been claimed that wolves, dogs, and, indeed, all mammals other than humans and some of the other primates are color blind. But is this description of the sight of the majority of mammals reasonable? If it is, one wonders why colors have developed in the natural world and why they are so important for camouflage? Why, for instance, does a deer fawn hide in shrubbery,

"Wolves are frequently met with in the countries West of Hudson's Bay, both on the barren grounds and among the woods, but they are not numerous; it is very uncommon to see more than three or four of them in a herd. Those that keep to the Westward, among the woods, are generally of the usual colour, but the greatest part of those that are killed by the Esquimaux are perfectly white."

SAMUEL HEARNE, ENGLISH EXPLORER AND FUR TRADER, FROM *A JOURNEY FROM PRINCE OF WALES'S FORT IN HUDSON'S BAY TO THE NORTHERN OCEAN, 1769-1772*

A lone wolf in open grassland.
© Thomas Kitchin

"concealed" by its coat? In a world where animals only saw objects in blacks, grays, and whites, there would be little need for the intricate variety of colors and tones that are to be found in any part of the world where wolves continue to exist, or, indeed, have ever existed. For these reasons, I believe that wolves do see color, but it is unlikely that they see the various hues of the spectrum as humans see them because the physical makeup of their eyes is different.

We frequently assume that the natural environment that we enjoy and study is the only one worth considering. In terms of color, however, many experiments have shown that different animals see the colors in the various parts of the light spectrum in different ways. We, for instance, can see the light of wavelengths that vary from 4,000 to 7,200 angstroms, but some animals — bees, for example — see wavelengths that vary from 3,000 to 6,500 angstroms. One researcher did find that dogs could distinguish red, yellow, blue, and green. To date, similar studies of wolves have not been published, but I have noted empirically that they seem to be stimulated by color changes, especially by red, orange, and yellow.

All in all, when the modern-day wolf emerged to take its place in the natural world, it arrived exceptionally well designed and suited to the role that it was meant to play, which is that of controller of prey species, animals that, without *Canis lupus*, would produce more young than the environment could sustain. Indeed, during the past century, man's depredation of the wolf, as well as the environment as a whole, has created habitats in which certain animals, especially members of the deer family, are free to overbreed. The result in many regions of North America and Eurasia has been the destruction of regional forests and the weakening of natural ecosystems as a whole.

Chapter Two

THE PACK

In Scotland, the term *pack* referred originally to a group of people who enjoyed close friendship. In this context, it is certainly well applied to wolves. Their social behavior resembles that of a well-ordered human family. Wolves are more social, in fact, than the primates, which are the most human-like mammals physiologically, but which have developed social systems that differ widely from those of *Homo sapiens.*

Indeed, early humans may have learned to live in closely related family groups after observing the efficient, cooperative, and highly social behavior of the wild dogs. The basis for this thesis is rooted in anthropological evidence which indicates that when early hominids evolved some three million years ago, other mammals had already become well established one to two million years earlier during the late Miocene epoch. Fossil discoveries confirm that the oldest known hominid mammal, which has been labelled *Australopithecus afarensis*, appeared at a time when wolves and other wild canids had already become efficient hunters of large prey.

At that time, it has been suggested, the small hominids were existing on a largely vegetarian diet, but were almost certainly also foraging for frogs and insect grubs and, perhaps, managing to catch

This fight between two more or less equal adversaries may have begun over a piece of the kill.
© John and Ann Mahan

small mammals, such as mice and voles. In addition, although there is no hard evidence to prove it, one can almost be assured that they made use of the remains of prey killed by carnivores. While following the hunting packs so as to be on hand when the predators became sated and abandoned their kills, early hominids may have realized the value of a close-knit social system.

In any event, field observers who seek to understand the behavior of wolves soon discover that they have set for themselves a difficult task, not only because the animals move about a great deal, often travel silently, and are quick to detect human presence, but also because every wolf is an individual. Physiologically, no two wolves look exactly alike. They may be of the same size and of similar coloration, but there are always differences, especially in their faces, which are readily recognized by an observer who has had close and relatively long contact with a pack. And, outside of a pack's social rituals, no two wolves behave in the same way. They walk differently, they play differently, they adopt different postures when they are drinking, their voices are different, and they howl differently.

Pack Size and Organization

The size of a wolf pack is extremely variable. The average pack in the northern parts of North America and Eurasia probably contains between six and nine wolves, although in the far north packs of up to thirty-six animals have been reported, but not confirmed. Conversely, packs that have become decimated by disease, human hunting, poisoning, trapping, and the destruction of habitat leading to a scarcity of prey may consist of only two or three animals until

The wolf at right is showing submission to a higher ranking individual.
© Thomas Kitchin

they can again build up their numbers, depending on whether or not there is a breeding pair in the group. However, if their habitat has been significantly decimated, and if human pressure continues, such wolves may die.

A wolf pack is led by its two founders, which are referred to as the alpha pair. Such a unit can be viewed as a cooperative family. At its head are the father and the mother, followed, in descending order of rank, by other adult wolves and the current year's young. In an old, established pack, some of the adults will be siblings, while others may be genetically related wolves from a neighboring pack.

Discipline is maintained by the parents, or if the founders of the pack have died, by the next high-ranking male and female. Order in the pack is achieved by various postures, stares, and, occasionally, by some bloodless physical punishment. There are times when wolves that rank below the alpha pair will engage in status interactions which may well end in bloodshed or even in the death of one of the combatants. Mortal conflicts are not common, however, because competition for status begins and is generally resolved early in the lives of the pups.

Nevertheless, if one or the other of the leaders shows weakness, becomes injured, or has been slowed by age, the next dominant, or beta, wolf will take over. Sometimes this change of leadership occurs after a noisy and bloody fight, which may end in the death of the aging leader. On other occasions, it is accomplished through sheer intimidation or harassment. By and large, though, the wolves of a healthy pack inhabiting a territory where prey is sufficient for the needs of all, lead generally peaceful lives. Their strong hierarchical ties largely dissuade pack members from engaging in negative behavior. In

Wolf communication: tail positions.

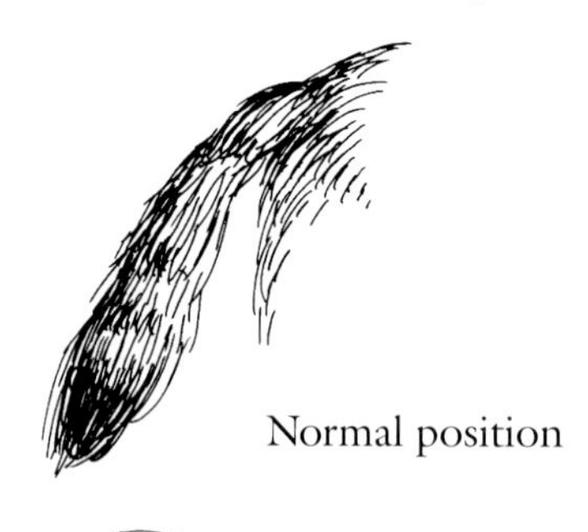

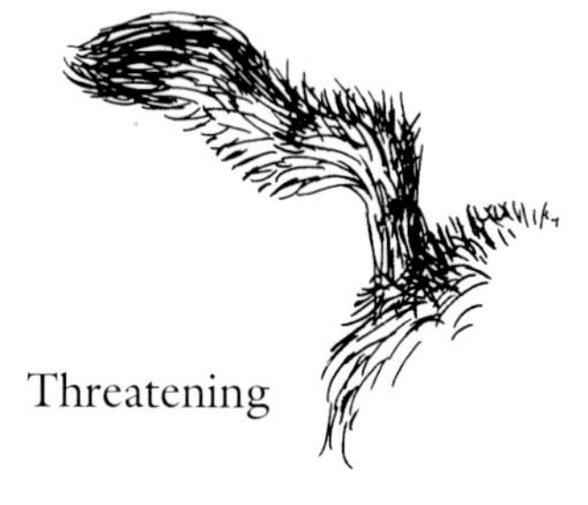

fact, wolves show remarkable affection for one another and generally "kiss and make up" following one of their brief altercations.

Wolves advertise their status by the ways in which they carry themselves. The leaders, during normal family interactions, walk with heads up, tails partly erect, and eyes fixed on any pack members within their orbit. Their demeanor is clearly relaxed. They *look* confident, assured of themselves and of their status. Lower-ranking wolves, on the other hand, always show deference when approached by a wolf of higher standing. Usually, their tails are held low, almost between their legs, and their ears are folded backwards. They crouch slightly, causing the back to arch, and their tongues flick in and out of their slightly opened mouths, preparing to lick their superiors.

On other occasions, a lower-ranking wolf will drop on its side, tail tucked between its legs, and lift one leg to expose the groin area. At the same time it will arch its head backwards to bare its throat, a submissive "grin" on its face. The action of deliberately exposing the two most vulnerable parts of the body is accepted by the dominant wolf as a sign of submission and reaffirms the higher-ranking wolf's dominance. However, if a pack member adopting such a posture is viewed by the alpha male or the alpha female as a threat to its leadership, the underling probably will be attacked. Should this occur, the subordinate wolf will probably squirm away, then jump to its feet. Tail tucked between its legs and voicing high-pitched, puppy-like sounds, it will seek to escape, running at full speed and pursued by the dominant animal, which may be joined by the other alpha as well as by subordinate pack members. Outside of the breeding season, such altercations are usually short-lived. The fleeing wolf will hide for a time before returning cautiously and clearly submissively to the

Wolf communication: facial expressions.

Normal expression

Apologetic

Happy

Warning

pack. All the members, led by the two leaders, will then greet the outcast with friendship and much tail wagging.

Usually, if a subordinate is committing a trespass, wolf leaders need only to stare fixedly at the offender in order to make it behave. The stare will cause it to give ground or even abandon a piece of meat or a bone. When a relaxed dominant wolf, whether standing or lying down, is disturbed for any reason, it becomes immediately aroused. It will stand stiff-legged, tail partly raised, ears pricked forward, and eyes fixed upon the cause of its arousal. Should its disturbance be caused by some serious event, such as a quarrel between two lower-ranking wolves or a clear act of trespass by a subordinate, the leader's hackles will become erect and the guard hairs will rise all over its body, from the tip of the tail to the top of the head. In such a pose, the wolf looks rather like a gigantic bottle brush! At the same time, emitting a loud, sonorous growl, it will peel back its lips and open its mouth part-way to reveal the large and lethal fangs. Confronted in this way, a subordinate wolf will immediately become submissive by flopping down and making itself easily vulnerable.

For the most part, though, a wolf pack is a highly social, gregarious unit. Its members almost invariably travel together, whether they are hunting or just moving from one part of their range to another. In the latter situation, led by the alphas, they usually trot one behind the other; in fact, they tend to travel so closely that each wolf steps in, or very close to, the tracks left by the leader. When a pack is travelling through snow, an uninitiated human observer might conclude that only one wolf had left the tracks.

Why do wolves travel in this fashion? Ease of travel may be a factor when moving through snow. It is easier to travel along a

Wolves howl standing, sitting, or lying down. © John and Ann Mahan

broken trail. In other situations, they may need one another's nearness for reassurance, or they may simply enjoy close contact. From my own observations, I believe that the latter is probably true, for wolves, despite engaging in some serious altercations, are inherently friendly animals. A pack leader or high-ranking wolf often shows friendly dominance by closing its teeth on a subordinate's muzzle in a ritual bite. The bite does not draw blood and is, indeed, as much a sign of affection as of benign discipline that reinforces the status of each of the participants. At such times, a subordinate responds by tucking its tail between its legs, lowering its head, and whining, in some instances also wetting itself submissively. Such behavior can occur at almost any time and for no apparent reason.

Similarly, wolves often howl just because they feel like doing so. While resting, one wolf may decide to howl. Seconds later all the other members of the pack begin to sing as well, some of them standing, others sitting, and yet others remaining stretched out on the ground. I have observed wolves engaging in this kind of howling many times, and it has always made me think that, somewhat like humans, they enjoy a singsong.

Wolves also howl in a ritual fashion before a hunt. Subordinates will mob the leaders for a few moments before the entire pack bursts into song, usually clustered together with heads raised and mouths agape as they launch their melodious yet primal voices.

In recent years, it has been claimed that wolves modulate their voices to fool neighboring packs into believing there are more wolves in the pack than there actually are. It has also been suggested that pups in a rendezvous deliberately modulate their voices to fool alien packs into believing that adults are present with the young. I cannot

Opposite: *These wolves are howling in unison, perhaps because they are about to go hunting.* © Karen Hollett

accept such thinking. It is difficult to fool a wolf, and in any event, such theories do not take into account the fact that in many instances packs relate with one another in a neighborly manner. In fact, I believe that wolves howl at times to communicate with their neighbors, probably doing so in order to notify their whereabouts to one another. As for wolves replying to human howling, I doubt that they believe they are responding to other wolves.

Although it is true that wolves can alter their voices considerably, I believe this occurs because of individual caprice. One female captive wolf howls beautifully when she wants to, but often flicks her tongue up and down in her mouth as she "sings," making an atrocious racket. She usually resorts to that kind of howling when she is excited; for example, at the approach of people bringing food. Every wolf has its own way of howling. When a person is familiar with the calls of a particular wolf, its voice can be quickly recognized, much as one can identify the voice of a friend or family member.

Wolves at Play

Wolves like to play. At times a pack member will clearly signal its desire to play by raising its tail slightly, pricking its ears forward, and bowing to a companion while raising a front paw, as though to wipe its muzzle. Then it crouches. If the companion wants to play, it will bound forward.

During such antics, the initiator may remain crouched, or it may rise and dash away. In the former situation, the approaching wolf will jump on the other wolf and the two will roll over and over, emitting mock growls and giving each other ritual bites. If one

A bonded pair engage in an affectionate greeting.
© Thomas Kitchin

animal jumps up and runs, the other will chase it, and the two animals will dash around in fast circles, jumping over obstacles and perhaps barking shrilly, or yapping much like a domestic dog. When two wolves engage in such play, other pack members, including the leaders, are likely to join in the game.

Early one morning in winter some twenty-five years ago, at a time when I was living in a wilderness cabin in the Ontario backwoods, I awakened to the excited calls of wolves. Their voices were coming from the direction of a large beaver pond. Because the calls continued from the same general location for several minutes, I concluded that a wolf pack was playing on the snow-coated ice of the pond.

I had been studying that particular pack for three years when, early one morning in spring, I was fortunate enough to find the den. It took weeks of patience and effort before the pack accepted me as a neutral, but after evidently deciding that I was harmless, they lost much of their caution when I was out and about and became curious about me. They often tracked and followed me when I was trying to follow them.

The following year, while the alpha female occupied the same den, I was able to watch the pack leader, a large, gray-fawn-and-white animal, and four of his subordinates, from a distance of approximately 200 feet (60 m). About an hour after I had taken up my station, the dark gray alpha female emerged from the den with her two small pups. The size and movements of the little whelps suggested that they were about three weeks old. They were both males and I christened them Rom and Rem.

Like all of their kind at that age, the pups were clumsily active little creatures of a brownish color with pug faces and small ears.

These arctic wolves are enjoying one another's company.
© J.D. Taylor

Under the watchful eyes of their mother, they spent time romping just outside the den entrance, and although still wobbly, they were always quick to dash into the tunnel upon hearing the slightest alien sound.

By the following January, the two pups were almost full-grown and had become useful pack members. However, the day before I heard the yapping, howling, and barking coming from the direction of the beaver pond, I had not seen the wolves for six days, nor had I heard their calls. I had almost concluded that they had left the region when I heard the lupine voices. Were they *my* pack I wondered, while I dressed hastily, but warmly, for –40° weather?

As I walked through the snow, making as little noise as possible, I continued to hear their calls coming from the same general area. Ten minutes later, moving through heavy brush for concealment, I reached the western edge of the pond and saw the pack. They were indeed playing, and boisterously!

Crouched about 500 feet (150 m) from the wolves, I was able to watch their movements through field glasses. The performance, I thought, was a lupine version of a child's game known in England as "hare and hounds," in which one individual is the hare, and the remainder are the hounds who seek to catch the quarry. The wolves played it at high speed and with great élan, at times howl-barking, on other occasions yelping shrilly, and now and then crying in pain when inadvertently hurt through collision. Collisions occurred quite often on the slippery ice surface that lay concealed, for the most part, by a shallow layer of soft snow.

The alpha male was the "hare," I noted. He was being chased by his mate and the other six wolves that now formed the pack. At one point, he was about three lengths ahead of the second-ranking male when, with the grace of a ballerina, he swerved, sliding in a controlled way on the snow-covered ice. As he righted himself, he charged the beta male, bowling him over in the snow to the evident delight of the other members of the pack, who increased their calls and ran to the beta as he sprang to his feet. Meanwhile, the leader had stopped and was watching his family. Seconds later the chase began anew, and once again the alpha charged a pack member, knocking it down just as he had the beta.

After I had been watching the wolves for about six minutes, the speed of the chase and its erratic movements took the pack to the far side of the pond. There, they continued to play for some moments, but, quite suddenly, the alpha pair stopped, stared across the snow at my place of concealment, and bolted into the forest, followed instantly by the rest of the pack. Evidently they had caught my scent, even though the slight breeze that stirred the evergreens was not blowing from my direction.

I was puzzled by the behavior of the alphas. In the past, they had allowed me to observe them from a lesser distance, and they had always known that I was in their immediate neighborhood. On that frosty morning, however, I concluded that I had caught the pack by surprise, which is something that wolves appear to dislike intensely because, I believe, the leaders feel that they should be constantly aware of events in their environment. When surprised, their behavior suggests that they feel embarrassed or, perhaps, guilty. At any rate, the male and female of the East Pack, as I had labelled the wolves, certainly overreacted to my familiar scent.

The pack's acceptance of my presence had been clearly shown by an incident that occurred the previous autumn. I had been

These Mexican wolves, photographed in captivity in Arizona, look fierce, but they are showing affection as the male (left) *gives his mate a ritual, painless muzzle bite.*
© C. Allan Morgan

following the wolves and had been sleeping out in a down bag that was covered by a tarpaulin to keep out the night moisture. But I was awakened at dawn by what I at first thought was the sound of rain hitting the tarp. On raising my head, however, I was just in time to see the alpha male drop his right leg, then he stepped back, sniffed the wet he had deposited at the foot of my bed, and turned around so that he could scratch dirt on the tarp, some of which landed on my face. Behind him, watching the performance, was the pack, their faces reflecting pleasure, which wolves show by opening their mouths slightly and smiling — quite literally. I was the victim of a wolf prank, the kind of playful behavior that is frequently enjoyed by these animals.

My relationship with that pack was to last for five years, a time during which they taught me much about themselves and about wolves in general.

Wolf Territories

Long before humans adopted similar tactics, wolf packs worked as individual family units within which each member cooperated for the benefit of all. Each pack claimed its own territory, as wolves continue to do, hunting over a range that today can vary widely in size, depending on the abundance of prey and the composition of the landscape. In the mountain country of Canada, for example, where meandering valleys and passes furnish the most likely habitat for prey, wolf territories tend to be larger than in less mountainous regions where the population density for prey is higher and the terrain is less inhospitable. In addition, home ranges in summer tend to be smaller than in winter, and summer ranges tend to be smaller in forested regions than on the tundra, where wolves may venture up to twenty miles (35 km) from the den.

In Ontario, in 1960, provincial government biologists surveyed a one-thousand-square-mile (2600-km^2) forested region and found that it contained twenty wolf packs. Statistically, this showed that each pack hunted over fifty square miles (130 km^2) of territory, but tidy divisions of that kind rarely occur in the wolf's real world, where some packs can earn a good living in a relatively small region and others have to travel much longer distances before they can make a kill.

Biological studies further suggest that, on average, wolf numbers in a pack in North America and in the tundra/boreal regions of Eurasia will not exceed one wolf for every ten square miles (26 km^2) of territory. According to the above calculations, a pack occupying forty square miles (100 km^2) would number only four wolves, while its neighbor, which might occupy sixty square miles (150 km^2),

Two subordinate wolves show submission as they greet the pack leader. Peter McLeod/ First Light

would number six wolves. Here again, it must be emphasized that such neat calculations are seldom, if ever, found in natural systems.

Wolf territories, however, are never static. Following game trails, wolves probably enter new territory frequently. That territory may or may not be claimed by another pack, or a pack may visit one end of its range only once or twice each year. In any event, there are no true boundary markers because there are no rigid boundaries.

In the matter of territorial boundaries, much has been said about "scent posts," a term that applies equally to many objects that wolves have urinated against, be these tree stumps, rocks, brush, standing trees, or a particular spot on the forest floor where a wolf has simply stopped to urinate out of need. Wherever one wolf has urinated, others coming within scent of the place will stop, sniff intently, and also urinate. Urine marking is thought to be strongly related to social status. Usually, it is the alpha male that initiates the ceremony, after which he stands back from the spot and often kick-scratches vigorously with his back feet to leave even more of his scent on the ground. Female wolves appear to scent-mark less often than males.

For a long time, it has been claimed that such places are boundary markers kept up by a pack for the sole purpose of warning other wolves against trespass. To date, I have not come across any evidence that would confirm such a belief, which I think stems from our own rigid concept of a territory as a place that is constantly guarded by troops or police against entry by unauthorized individuals. Instead, from my own field studies of wolves, I tend to agree with the German biologist R. Shenkel, who noted that scent-marking represents a nonaggressive kind of "contact between neighbors."

Wolves in Montana smell the snow around a clump of bushes. If the scent of alien wolves is detected, they will cover it with their urine, then scratch the spot with their back feet.
© Robert Zakrison

It has also been claimed that if a human urinates on a so-called scent post, wolves will hurry to the spot and cover the human scent with their own. For my own satisfaction, I have tried the stunt many times without being able to arouse any but a momentary, clearly disinterested sniff at the place that I have marked. On the other hand, wolves will scent-mark on all kinds of other urine and fecal deposits if they pick up the odor of other wolves, coyotes, skunks, raccoons, moose, deer, and so on — the potency of the odor that they have covered telling them whether the deposit is recent or old.

A few years ago, when seeking a suitable Christmas tree on our forest property, I found a large balsam fir that had crashed because of fungal disease in its root system. The top of the tree was still alive and green and ideal for my purpose, so I knelt in the snow and cut it off, noticing as I worked that some of the branches and needles had adhering to them a number of small, yellow globules of what I took to be balsam gum.

I shouldered the tree, took it home, fixed it in its container, and then placed it in front of one of our living room windows. Soon afterwards my wife, Sharon, began to decorate the tree, but after about ten minutes she turned to me and said that there was "a funny smell" coming from the branches. I assumed she meant the scent of balsam resin, but I went to investigate. One sniff told me that I was smelling wolf urine. Almost at the same moment, I noticed that the yellow globules that I had mistaken for balsam gum had melted! Sharon, accustomed to such minor surprises, finished decorating the tree, and the somewhat musky odor of lupine urine did not take too long to evaporate in the heated house. While it was clearing, I returned to the fallen tree to note, now that I was looking for the signs, that one or more wolves had aimed their streams over the end of the balsam. I also noted what I had missed earlier — an area of yellowish snow in the place where the "Christmas tree" had rested.

I had been in that particular place just four days earlier, before the balsam had fallen, and although I had seen wolf tracks in the immediate neighborhood and had found two places some distance away where the pack had stopped to leave its communal scent, there had been neither tracks nor urine stains in the place where the tree had landed. Now, checking the snow carefully, I noticed one set of wolf tracks leading up to where the snow was stained, then moving away through heavy underbrush. I backtracked the spoor and soon found the spot where one wolf had left the pack to urinate on the tree. Returning to the urine stains, I followed the tracks, which turned away from the balsam and headed at an angle to a location thirty strides away, where the animal had rejoined the pack.

Seeking to determine whether or not that particular location was a scent station, I continued to visit it throughout the winter. I found a number of places in the general neighborhood where the visiting pack had urinated, but none of the wolves had returned to the downed balsam. Apart from that experience, I have over the years examined many locations where wolves had urinated on a variety of objects. In a number of areas they had urinated on trees, stumps, or brush in places that were only five or ten paces away from other marked locations.

Why, then, do wolves and dogs scent-mark? One reason may be to advertise their presence, perhaps as a greeting to other wolves. It may also be that because scent means so much to all wild animals, and especially to wolves, the odor of one animal stimulates another to add its own scent to the place, as if to say, "I have been here also." In addition, it certainly alerts other packs to the fact that they have entered a region already occupied by a wolf family, information that may or may not cause the "intruders" to leave the area. And, if for no other reason, scent-marking allows for kin recognition.

Wolves often engage in aggressive behavior that usually ends without injury.
© John and Ann Mahan

Relationships between Packs

In October, 1991, at the request of the Ontario Society for the Prevention of Cruelty to Animals, we agreed to care for a mistreated four-month-old wolf, which had been the "pet" of a motorcycle gang. When the wolf, which we named Silva, had been with us for about three months, we were visited by friends, Bob and Jackie Gurr, who have a Siberian husky. Jackie was introduced to the wolf, who accepted her presence, if not her touch, in the enclosure. Bob kept away, for Silva had a great fear of men because of her past experiences with them. We wondered, though, how she would react to the Gurr's husky, so Jackie led Halley to the fence. Silva went into a panic!

Later, I was lead-walking Alba, one of our captive wolves, and she

At this pack greeting ceremony, subordinates mob one of their leaders to demonstrate affection and unity. © John and Ann Mahan

headed for Silva's enclosure. At first I felt that I should not allow Alba to get near the young wolf, but, watching Silva, I realized that she was expressing interest rather than fear, so I allowed Alba to walk up to the wire. To my surprise, Silva came right up to the fence, tail wagging vigorously, and whining. Her behavior was that of a young wolf towards a senior member of the pack. Silva was so anxious to make contact with Alba that she actually forced her head through the four-inch by six-inch (10–cm by 15–cm) wire and licked the other wolf.

The next day, the Gurrs arrived to show us some of the photographs that Bob had taken of both Alba and Silva. When we told them about the small wolf's reaction to the mature female, Bob and Jackie wondered if Silva would now also accept Halley. I thought it was a good idea to try again, so I asked Jackie to lead the husky to the enclosure. Before the dog got near the wire, Silva became extremely agitated.

I was delighted, not because Silva had become frightened of the dog, of course, but because her reactions indicated that wolves are capable of detecting blood-lines or genetic relationships. Clearly, Silva had instantly recognized Alba as an animal of her own kind; just as clearly she had recognized the dog as an alien. Years earlier, after observing many wild packs, I had become puzzled by a controversy that exists to this day among wolf biologists. On the one hand, some state categorically that if one pack "trespasses" on the territory of another, the homeowners will fight the invaders, and certainly, there is field evidence of such battles, some of which have led to the death of wolves. On the other hand, there are biologists who have evidence that wolves are not aggressive under such circumstances. In my own case, I have seen examples of both reactions.

On three separate occasions in different regions of Canada, I have observed packs enter the territories of their neighbors. In each case, the two alpha males, watched expectantly by their companions, approached each other, tails erect and hackles raised. They walked stiff-legged, but slowly, towards each other, and when they were almost nose to nose, both began to wag their tails. Seconds later, the two packs mingled and began to play, gamboling around the small clearing where the meeting had taken place. They jumped over each other, knocking one another to the ground, and then raced around in circles.

While one wolf avoids conflict, two other wolves display controlled aggression, each demonstrating aggressive-submissive behavior at the same time.
© E.A. James/NHPA

On other occasions, however, the opposite had taken place and a battle followed. In one instance, one of the invading wolves, a young male, as nearly as I could judge through my field glasses, was badly injured before the invading pack ran away, hotly pursued by the territory holders.

Why had there been such different reactions? At first I thought that perhaps it was a matter of food. If there was plenty of prey for all, there may have been nothing to gain by fighting and, in fact, a great deal to lose. This seemed a reasonable answer; it still does. Yet, I feel there is more to it than that. A wolf study in Isle Royale National Park in Michigan promises to shed some light on the questions. The island is located in the northwest corner of Lake Superior and consists of 210 square miles (544 km^2) of rugged, well-wooded territory off the Ontario coast. Here, moose and wolves have been cohabiting since about 1951. Before that, after the mining and logging interests of the late nineteenth century had devastated the island's natural resources, a sparse population of moose continued to survive alongside beaver, red foxes, snowshoe hares, and a variety of smaller

When a high-ranking wolf approaches a subordinate, the latter demonstrates typical submissive body language.
© Pauline Brunner

mammals and birds. But there were no wolves left on the island.

Over the years, vegetation recolonized the devastated areas, offering more food for the moose. The moose population began to increase until, eventually, despite the rate of new vegetation growth, it increased to the point where the animals unwittingly began to create almost as much devastation as had the foresters. In a similar situation on the mainland, moose would have dispersed before their food base was seriously threatened, but on Isle Royale the big deer could not leave the island. As a result, the vegetation was largely destroyed, and the numbers of moose dropped dramatically.

In time, as is the way of nature when undisturbed by human activity, the vegetation began to recover and the moose, keeping pace with the food base, began to increase again. Biologists of the United States Fish and Wildlife Service in Washington, D.C., began to worry. If the moose continued to increase, would they once again devastate the island's vegetation? It seemed likely.

Then, in 1951, somebody found a wolf track on the island, made a cast of it, and sent it to Washington. The biologists were cautiously delighted, but puzzled. How had a wolf, or wolves, reached the island? It seemed likely that one or more wolves had migrated from Ontario during a bitterly cold winter that had laid down a heavy layer of ice over the lake. The animals had probably trotted over the ice — between fifteen and eighteen miles (24 and 29 km), depending on their point of departure from Ontario — to reach the island. Data relating to that crossing are unavailable, of course, so it is not known why wolves from Ontario would have found it necessary or expedient to brave the dangerous journey. In any event, inasmuch as wolf numbers increased slightly over the next three years, it

is certain that at least one male and one female settled on Isle Royale.

In 1957, biologist Dr. Durward L. Allen was instrumental in organizing an ongoing wolf-moose study on the island. He became the study's director and continued in that capacity until 1975, when Dr. Rolf Peterson became director.

The importance of the study lies in the fact that, perhaps for the first time, it has become possible to monitor closely in one isolated ecosystem that is now relatively undisturbed by humans, the predator-prey relationships and the effects that moose and wolves are having on the island environment. The information that has been emerging from the program shows that under conditions such as existed on the island during the early years of the study, the moose continued to increase despite predation by wolves. The predators were also increasing their own numbers in order to keep pace with the food base.

Thus, for a number of years, it seemed that moose and wolf populations might continue more or less in balance with the food base; that is to say, the numbers of both species would probably fluctuate, seesawing up and down, but neither multiplying dramatically nor dropping radically. It did not happen that way. In 1970 there were about 1,200 moose and some eighteen wolves sharing the island. Then, in 1971, the moose crashed, their numbers dropping to about 750 at a time when the number of wolves had increased to twenty. From that point on, numbers of both species fluctuated rapidly. The wolves became more numerous, but crashed in 1977, dropping from a high of about forty-two to about thirty-three, then recovering the following year and climbing steadily until they reached a peak of fifty animals in 1980. By this time, however, the

The wolf in the foreground shows dominance with bared teeth, and ears erect and pointed forward. The wolf in the background shows submission by flattening its ears and holding its head and tail down.
© Wm. Munoz

moose population had crashed to slightly more than 600 animals. A year later, as moose numbers began to increase, wolf numbers dropped. In Peterson's annual report for 1991/92, he notes that moose numbers have climbed to a high of 1,600, which is close to the highest number recorded during the past forty years. Wolf numbers, on the other hand, have dropped to twelve.

The major question being asked is whether wolf numbers, at least on Isle Royale, are regulated by their food base or by the negative effects of inbreeding. The former case can be expected to resolve itself in the near future as old moose increase in numbers and become easier prey for the wolves. The latter case may be harder to determine, inasmuch as the genetic make-up of the original wolf migrants is not known. Nevertheless, data already gathered and documented by Allen, Peterson, and others have thus far shown that food is, indeed, "the burning question in all animal societies," to quote the English biologist Charles Elton.

Should it be determined that the wolf decrease is caused by genetic deterioration, however, that too will offer important knowledge. I have field evidence that suggests that wolves may well be able to inbreed for a considerable period for the simple reason that genetically inadequate pups are unlikely to survive to adulthood, while genetically healthy siblings will grow up to take their place within the pack. If my field observations can be scientifically confirmed, I would expect to see an increase in the Isle Royale wolves by about 1993/94. Conversely, if genetic deterioration is the cause of the low numbers, it will probably cause the ultimate demise of all the island's wolves.

In any event, it would seem that genetic relationships between wolf packs may explain why some packs meet in friendship when

The backward-sloping ears of the subordinate indicate respectful submission.
© Karen Hollett

food is plentiful. On the other hand, if food is in short supply, territories are likely to be more jealously guarded and confrontations between related packs may well occur, just as on occasion they may occur within a family pack, and, in fact, among members of a human family. Indeed, there are more than enough examples of severe conflict over territory among human populations. By comparison, those that may occur among wolves are insignificant.

Discussing with lay visitors the ability of wolves to recognize genetic relationships, I am invariably asked: "How can they do that?" The simple answer is that, apart from having an incredibly acute sense of smell, genetic imprinting allows them to recognize the odor of their species and, it is evident, even the odor of closely related kin.

When our two captive wolves, Tundra and Taiga, were five months old, my daughter, Alison, came over from England on a visit. The pups had never met her, but when she accompanied me to their enclosure, I was surprised by their behavior. As we approached, they both reacted as they always do to my presence. It was as though I alone was walking towards them.

Although I was puzzled, I made no mention of it to Alison as I opened the enclosure's gate and ushered her ahead of me as we entered. To my further surprise, instead of running away from the "stranger" as they so far had done with all those first-time visitors who had been allowed inside the enclosure, they dashed up to Alison, mobbing her, as wolves will do with a family member who has been away for a few days. When she sat down on the step of the barn in which we feed them and where they shelter in summer from the heat and biting flies, Tundra and Taiga flopped at her feet.

Alison began to stroke them, one hand on each, but a few

moments later her stomach rumbled. At the sound, both pups tucked their tails between their legs and wet themselves in submission. They had mistaken the stomach sounds for growls, and they had responded to them just as they responded to me on those few occasions when I had needed to "scold" them by uttering an imitation, and not very good, growl. Obviously, they were treating my daughter as a related and senior alpha female!

I did not, however, talk to Alison about my conclusions, nor did I mention them to Sharon when we returned to the house. Instead, I suggested to my wife that it might be nice for Alison to lead-walk Taiga that afternoon. We usually took them out twice a day. Sharon walked Taiga as a rule, but when the young bitch pulled too hard for her, we would change over, and Sharon would walk Tundra, who was always easier to manage. Taiga, who saw me as the alpha male, would walk sedately and obediently knowing that it was my hand at the other end of the lead. How, I wondered, would she react to Alison? At three o'clock that afternoon I found out.

Taiga responded to Alison's control just as she responded to mine. It became clear to me that the young wolves had recognized our family relationship and, although our pheromones were not those of wolves, the pups nevertheless associated them with their upbringing and comfort.

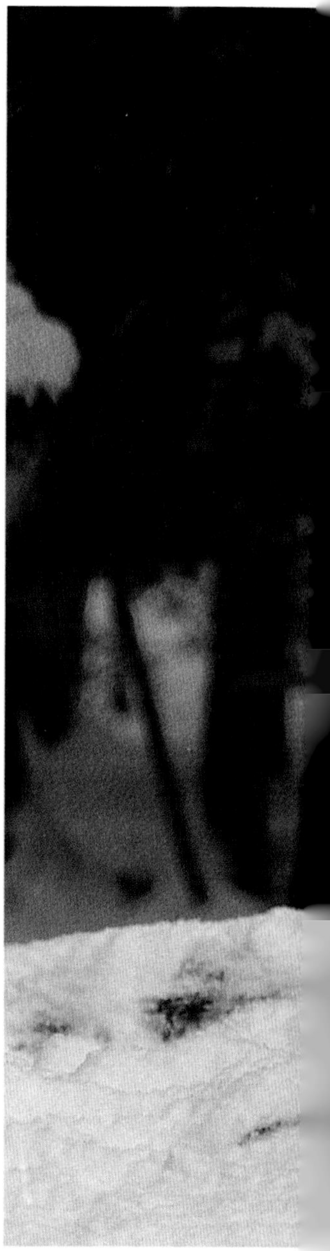

A low-ranking wolf displays submission to the male and female pack leaders.
© Wm. Munoz

Chapter Three

THE HUNT

Referring to the social interactions that frequently occur when wolves are about to set out on a hunt, the late Adolph Murie noted in his book, *The Wolves of Mount McKinley*, that a pack he had been observing for many days gathered one evening near the den in which the alpha female was still nursing that year's pups. They began to lick one another while wagging their tails, then "they all howled, and while they howled, the gray female galloped up from the den 100 yards (90 m) and joined them. She was greeted with energetic tail-wagging and general good feeling. Then the vigorous actions came to an end and five muzzles pointed skyward. Their howling floated across the tundra." When the howling ended, wrote Murie, the female returned to her den to care for the pups and the remaining four wolves set out on the hunt.

Such behavior, although it occurs fairly often, does not take place every time a pack sets out to hunt. Similar demonstrations of affection and excitement that I have witnessed have always occurred in the neighborhood of a den at a time when the mother has been caring for pups, so it may be that the arrival of young, always an exciting and joyous time for the pack, triggers the exuberant behavior. In any event, wolves also set out quietly to hunt after one

A cow moose and her half-grown calf. Wolves would be less likely to kill the calf while such a vigorous mother was present.
© Bob Gurr Photo

Hunting musk-oxen on Canada's barren grounds. A pack of arctic wolves (opposite top*) is temporarily stymied by the defensive tactics of a herd of musk-oxen, which has formed a tight line in front of the calves* (opposite bottom*).*

Soon, however, the wolves' maneuverings unnerve their prey (above*), the herd begins to panic, and the pack manages to bring down a calf* (right*).*
Sequence by David Mech

or both leaders rise from their resting places and trot away, immediately followed by the rest of the pack.

Usually wolves first detect animals by scent. Whenever possible, they seek to catch their prey on the run. Indeed, it has long been noted that wolves keep themselves fed with their feet, just as it is said that the wolf taught the deer how to run. This hunting method holds several benefits for mammals whose evolution specifically molded them for the chase. Because they are long-distance runners, their endurance is much greater than that of their prey. Consequently, during an extended chase, the quarry cannot keep the pace going for as long as can the hunters. Also, a fleeing animal is more vulnerable and less likely to injure a wolf than one that is standing at bay.

The beaver often feeds at the shoreline, where it is vulnerable to attack.
© Thomas Kitchin

Those prey animals that do stand at bay when approached by a wolf pack are likely to survive. Moose, elk, caribou, and even deer and mountain sheep have been known to repel wolves by simply refusing to run and by exhibiting aggressive behavior. Under such conditions, wolves will not usually initiate an attack. Instead, a pack will surround the animal, but the wolves will stay far enough away to avoid a sudden charge. Mountain sheep, for example, can butt with enormous force, and they are extremely agile, able to turn swiftly in any direction by rising on their back legs.

When brought to bay, members of the deer family can defend themselves with their pointed front hoofs, charging at an attacker and jumping upwards on their back legs while stabbing forward and down with their lethal front hoofs. If such stabbing blows connect, they can easily go right through a wolf or other predator. Similarly, if a wolf tries to approach a standing moose or elk from the rear while other wolves are approaching from the front, the intended victim is

A lone wolf is attracted to the scent of a beaver lodge. Inside, the occupants are well protected by the sticks and mud of their house.
© Thomas Kitchin

likely to kick backwards with its big rear hoofs. A connecting blow can seriously injure or kill a wolf. This means of defense is aided by the fact that the eyes of deer, elk, and moose are set almost on the sides of their faces. This position gives them relatively good backwards vision, whereas the eyes of the hunters are set closer to the nose, looking forward. For this reason, although a wolf must turn its head to look backwards, a member of the deer family can see to the rear relatively well without turning its head.

A wolf pack often lies down around a standing quarry and, now and then, one or two of the leaders will rise, test the animal, and retreat if it refuses to run or shows signs of aggression. When that occurs, other pack members usually walk closer and behind the animal. But provided that the quarry does not panic and run, the pack will eventually move away to seek easier prey.

This means of defense is also effective for smaller prey, as I witnessed in Ontario during a summer morning in 1965. I had been sitting quietly, watching from a distance of about 75 feet (25 m) as a large, male raccoon scrambled down a pine tree and began to eat a succulent, soccer-ball-sized puffball mushroom. The raccoon was fully occupied with its feast, and I was as fully engrossed watching him. Neither one of us heard the approach of two yearling wolves, animals born the previous spring that probably weighed 55 to 60 pounds (25 to 27 kg). The raccoon became aware of the wolves before they had entered my field of vision, but his behavior alerted me to the fact that something had frightened him, for he abandoned the puffball and ran to the pine. Before he could start climbing, however, the wolves dashed at him.

In an instant, snarling ferociously and showing his formidable

Like dogs, wolves when sated sometimes cache pieces of meat or bone to be eaten later.
Peter McLeod/
First Light

fangs, the raccoon turned around, backed himself up against the tree trunk and stood at bay. Although the young wolves tried several times to flush the quarry, the raccoon stayed where he was, continuing to snarl and to threaten one or the other with his fangs while emitting growls that seemed too loud and ferocious to be coming from such a small animal. The wolves continued to threaten the raccoon for about six minutes, after which they lost interest and loped away. The raccoon immediately relaxed, stopped growling, and returned to his feed of puffball. He finished the meal three minutes later and started to climb the tree.

This wolf has killed a wild goose. Lone wolves must make do with smaller prey, because they are less able to bring down large animals on their own.
© Thomas Kitchin

"Will he hurry if I stand up and move towards him?" I wondered. Moments later I had my answer. The raccoon, which was already about ten feet (3 m) from the ground when I rose and walked towards him, lashed his tail, wet himself, and streaked up the tree, not stopping until he was a good thirty feet (10 m) above my head. It was intriguing to note that when caught by surprise on the ground by the wolves, the animal had immediately become aggressive, but when surprised at a moment when he was beyond my reach, he reacted fearfully. In the one situation, the raccoon had no means of escape, so he used aggression in order to intimidate the wolves; in the other situation, once he was safe, he was able to give vent to the fear that had earlier been overridden by the need to defend himself.

A Wolf's Diet

Wolves are often accused of being vicious, greedy killers that overeat to the point of torpor. Nothing could be further from the truth! These wild hunters do, of course, eat meat, and lots of it. However,

After a large meal, contented wolves often howl.
© Thomas Kitchin

although they will gorge if they have gone hungry for several days or, as sometimes happens, for an entire week, they do not overeat. In fact, their ingestion of food is regulated by the liver, which stores excess glucose as glycogen, and by the hunger and satiety centers located in the brain. When the liver is topped up with glycogen, a wolf's satiety center is activated and it will not eat. After going hungry for a time, the liver's supply of glycogen is released into the blood as glucose. When the supply is depleted, the hunger center is activated.

The wolf's digestive system is so strong that it usually breaks down every bit of protein that has been eaten. As a result, wolf scats, especially between autumn and summer, contain very little, if any, fecal matter. The stool, which may be an inch (2.5 cm) in diameter and three or four inches (7 or 10 cm) long, is usually gray or white and contains chips of bone and fur compacted and held together by mucus. If such a stool is stepped on or broken up, its interior is yellowish and granular and, even in a fresh dropping, hardly any odor reaches the human nose, though other wolves can detect the odor. (Scats should not be examined too closely because they can contain parasite eggs, which can be transmitted to humans.)

Although wolves cannot survive without meat, they, like most other carnivores, will at times eat vegetable matter, especially tender grasses and herbs, and in spring they will gnaw at the bark of sapling trees. They will also eat fruit in season, and I once watched a wolf consume a number of mushrooms. Nevertheless, their principal food does, of course, consist of the meat of large prey species when it is available.

In North America, the main prey of wolves includes bison (now only in Wood Buffalo National Park, located partly in the Northwest Territories and partly in northern Alberta), moose, caribou, elk, deer, wild goats, and wild sheep. They also hunt beaver, porcupines, hares, rabbits, snakes, and birds, such as grouse, ducks, and geese. Pups, and adults looking for snacks, eat mice and voles. In fact, wolves will hunt any kind of prey that may exist in the many different regions of the world in which these widespread mammals are to be found. In the absence of their natural prey, wolves turn to domestic stock, causing farmers to call for their extermination.

Although wolves will kill an occasional animal in its prime, most of their prey consist of the old, the unfit, or the young. However, it is said with truth that when the wolf eats, everybody eats, for there are a great many lesser animals that benefit from wolf leftovers. These include wolverines, lynx, bobcats, mink, weasels, hares, porcupines, squirrels, mice, voles, shrews, and the ever-present ravens, which seem to have developed a close relationship with wolves everywhere and rarely fail to turn up whenever a pack has made a kill.

At times, ravens may lead wolves to an aging or sick moose, elk, or deer by flying over its location and calling raucously until wolves respond and make the kill. There are those who would argue that the ravens deliberately scout for the wolves, but it seems more logical to

Ravens routinely follow wolf packs and share in their kills. Only rarely will a young wolf try to kill one of the large birds.
© Henry Ausloos/NHPA

conclude that the big black birds, which are supreme opportunists, hang around a sick animal so as to be on hand when it dies — whether its demise results from old age, disease, injury, or from the teeth of wolves that have heard the ravens and are attracted to the locale.

It should not be supposed that wolves eat well and regularly. Like all predators, they go from feast to famine on a regular basis. In fact, widespread data regarding hunting by wolves in North America show success rates of between 7 and 10 percent.

In order to keep themselves fed, wolves must travel many miles and endure frequent famines. For these reasons, the two pack leaders are the first to eat after a kill has been made. This priority is prompted not by greed, but by genetic inheritance. The pack leaders are responsible for the well-being of the entire pack; they are the most active, the most daring, and usually the first to initiate an attack. After a time, however, the beta wolves are allowed to share the meal and, later, in descending order of rank, the other wolves will eat. If the kill is large — a moose or caribou, for example — there is usually more than enough food for all the pack members. But if the prey is small, some of the lower-ranking wolves may have to make do with scraps until the next hunt takes place. In the meantime, they must try to satisfy their hunger with small prey, such as hares, mice, and voles. Because they start out hungry and have to pounce on a lot of empty spaces before they can catch one small animal, they must work hard for these morsels of food.

As for the wolf's predators, man, undoubtedly, has been and continues to be its greatest foe. Occasionally, bears prey on wolf pups, but these attacks are relatively insignificant. Aside from humans, disease is this animal's most threatening enemy.

Chapter Four

FAMILY LIFE

Despite the fact that during the past millennia the human species has sought to remove itself from its ancestral roots, the basic beginnings of our own courtships and subsequent unions are not too different from those that take place during the formation of a new wolf pack, which begins when a male wolf roaming alone in search of a partner meets a female wolf that has wandered away on a similar quest. The plot is familiar: boy meets girl, they like each other, they court and marry, and they raise a family.

In the world of the wolf, such a meeting occurs during the breeding season. It may take place because the two scent each other over a relatively long distance, or the pair may meet accidentally. But most often, it results after one or the other wolf stops to signal its lonely presence by howling.

If the calls elicit a reply, the two animals, which may be a mile or more apart, trot towards each other, pausing now and then to howl. In this context, if the first howls emitted by a male are responded to by another male (or, conversely, should a female respond to another female), the sex odor of each will be perceived when the animals come within scenting distance. In that case, a meeting between the two is less likely to take place, although my field observations suggest

After mating, pairs continue to be affectionate.
© Wm. Munoz

that now and then two lone wolves of the same sex may join forces, for *Canis lupus* is an inherently social animal that does not survive well on its own.

In whatever way the contact occurs between two wolves of opposite sexes, however, and always providing that each wolf finds the other agreeably compatible, a time of active courtship will commence.

An affectionate scene repeated often by a courting pair.
© Erwin and Peggy Bauer

At first, tails wagging and rumps wriggling, the two approach each other while emitting short whines and little excited snapping sounds, mouths open and tongues flicking in and out, their lips peeled back in lupine grins, and their nostrils flaring as they investigate their respective scents. Then, they may touch noses, or they may lick each other in the mouth. Afterwards, as the female's hormones advertise her interest, the male will probably seek to initiate mating immediately, although at this stage he may not be reproductively ready, for, like the females, male wolves are sexually active only once each year. Except during the breeding season, a male's gonads are small, but when stimulated through the endocrine system by the scent of a receptive female, they start to increase in size, in due course swelling to about three times their normal mass.

A male's premature advances, therefore, are repulsed noisily, not only because the female can scent the unreadiness of her partner in the relatively weak odor of the testosterone hormone that is issuing from him, but also because she herself will not be sexually receptive until after her estrus blood begins to spot the ground. Thus, for a time that may last a few days or perhaps several weeks depending on when the female began to be receptive, the two roam the area, hunting small prey. They often stop to play; they may chase

A breeding pair is approached by the pack's second-ranking male.
© Karen Hollett

each other, or one of them may pick up a stick or bone and toss it in the air, after which both will dash to capture it. Between bouts of play, the hunting of prey, feeding, and resting, the pair will wander in search of a territory that is not already occupied by another wolf pack, their extraordinary ability to scent over long distances, as well as their keen ears, alerting them if they are about to intrude on an unrelated and already established pack. If they come across an alien pack, the pair will alter their line of travel.

In the far north, most matings take place in late March or early April; but in the more southerly regions of wolf range, they occur in early February or even earlier. In the midranges — in central Canada and the northern parts of the continental United States, as well as in similar latitudes in Eurasia and Asia — mating generally occurs between late February and mid-March. As a rule, wolves do not become sexually receptive until their second year, but now and then yearling wolves have been known to mate and breed, although in such cases the union occurs between a yearling female and an older male.

In an undisturbed pack of more than two wolves, only the leaders will usually mate. If a subordinate female shows signs of being sexually receptive, the alpha female represses the urge through what is, in fact, psychological intimidation. This may cause the subordinate to interrupt her estrus cycle through excessive discharge of adrenalin or may intimidate her to the point where, although she is in estrus, she will not seek the attention of a subordinate male. Subordinate males are in turn inhibited by the male leader from responding to a soliciting female. The rituals of pack order serve to promote birth control, keeping the pack's numbers in balance with the available food base.

In some instances, a subordinate wolf of either sex may be so determined to mate that it refuses to be intimidated by the leaders. In such cases, the subordinate wolf leaves the pack rather than submit to harassment, undoubtedly hoping to meet a partner of the opposite sex. If it does not encounter a partner, the loner will use its senses of smell and hearing to keep in touch with its packmates, often following the pack and eating leftovers from its kills. At the end of the breeding cycle, the loner will more than likely rejoin the pack if it has not found a mate. If it does, it will at first be ritually harassed by the leaders and the rest of the pack. Such treatment is bloodless and intended to show the bond that exists in the family as well as to put the wanderer in its hierarchical place.

In mating, the male canid's sexual tie, which ensures proper insemination, is much weaker in wolves than in dogs. Dogs usually

"They (wolves) always burrow underground to bring forth their young; and though it is natural to suppose them very fierce at those times, yet I have frequently seen the Indians go to their dens, and take out the young ones and play with them. I never knew a Northern Indian hurt one of them: on the contrary, they always put them carefully into the den again."

SAMUEL HEARNE, ENGLISH EXPLORER AND FUR TRADER, FROM *A JOURNEY FROM PRINCE OF WALES'S FORT IN HUDSON'S BAY TO THE NORTHERN OCEAN, 1769-1772*

have difficulty parting until the male has discharged all his sperm. In wolves, empirical observation suggests that the tie is weaker so as to allow a mating pair to part quickly in the face of an unexpected threat.

After mating has been completed, and if the pair has not already found a range in which to settle, the wolves will concentrate on the search for a location that offers abundant prey, that is well forested but yet offers some open landscapes, and, preferably, that contains rivers, lakes, or ponds. Ideally, the territory would also offer some high land that lends itself to the digging of a den and provides a good view of the surroundings.

As soon as the bonded pair have found a suitable territory, the female will start to dig the subterranean shelter, which is the preferred whelping nursery. But if the terrain is unsuitable for digging, she will have to make do with a cave or a hollow chamber among a jumble of rocks or, if nothing more suitable is available, she may seek the shelter of a brushy tangle of hawthorns, alders, or willows.

A den is located at the end of a tunnel, which can be thirty feet (10 m) long and may have two entrances. This subterranean approach may be more or less straight, or, if rocks are encountered, it may curve.

Digging a tunnel is a heavy task. The female, using her front paws, must thrust excavated earth backwards, under her own body, then kick it farther away with her back feet. As the tunnel gets longer, she will have to move backwards over the resulting mound and repeat her actions until she reaches the mouth of the tunnel. Here, she shoots out the dirt by again kicking it backwards. She then re-enters the tunnel, digs some more with her front feet, and repeats the same series of actions. Now and again, she will encounter plant

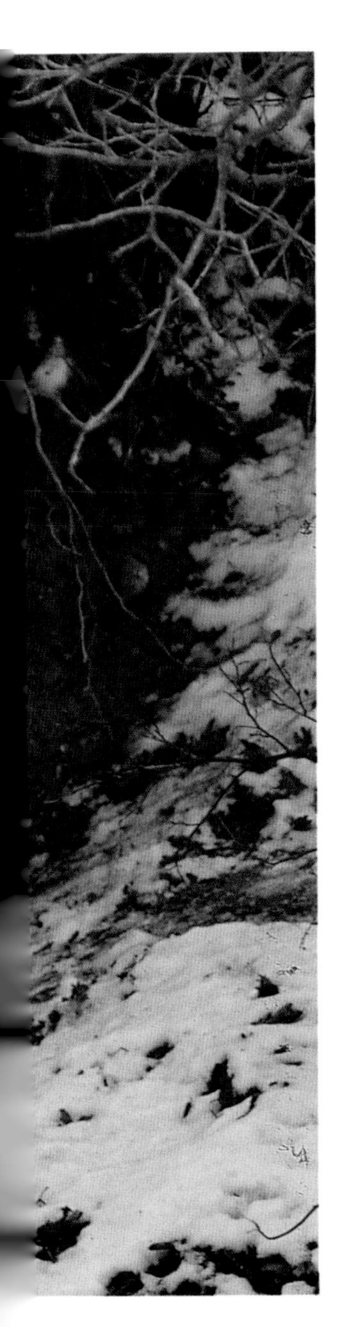

A female captive wolf at the entrance of her den.
© Denver A. Bryan

roots. The small ones she usually leaves hanging in the tunnel, but she must gnaw away any large roots.

The back-kick of a wolf is powered by the buttock and thigh muscles and, having been inadvertently struck by such a kick, I can testify to its power. Indeed, if one stands within eight or ten feet (2 or 3 m) of a den mouth while a wolf is inside, busy tossing out soil, the shower of particles, not to mention the small, flying stones, hits the bystander with an impressive force.

What causes a wolf to work so hard in order to dig a long tunnel when a short one might do just as well? The answer to this question may lie in the fact that wolves are inherently cautious. Provided that good soil conditions are encountered, the female will continue to dig until she reaches a point from which outside noises are muted. Conversely, the sounds made by the pups from that point will be less likely to reach the den entrance. The nesting chamber itself consists of a round, shallow basin scooped out of the bare earth. It is not lined with fur or any other bedding material, and it is usually located somewhat uphill from the mouth of the tunnel to avoid flooding.

It may be coincidental, but the absence of bedding litter ensures that neither the mother nor the pups will become infested by common fleas. These bloodsuckers use their hosts only as meal tickets and, when full, drop off in den litter to lay their eggs. The detritus of a lined and padded den invariably contains food for the emerging flea larvae. A wolf den, however, offers no food for the larvae. It is unlined and the mother keeps the nesting chamber clean by eating the fecal deposits of the pups. The ammonia-laden urine soaking into the ground acts as a further deterrent. Of course, wolves that mate in colder climes are less likely to have fleas, because these parasites are dormant under such weather conditions. In any event, wolves are not important hosts of the *Pulicidae* family, despite the fact that the scientific literature notes that they are hosts to some thirty species of fleas. This may be the case in some southern regions, or, more likely, among captive wolves living in unsanitary conditions, but I have not found fleas on wild wolves that have been live-trapped for tagging and neither have I observed fleas on captive wolves that have been kept in proper conditions, which would include a forested area and, if breeding is to take place, suitable den-digging soil.

A mother wolf picks up her young pup.
© John and Ann Mahan

Wolf Pups

North of 45° latitude, but with variations according to climate, most wolf puppies are born in late April or mid-May after a gestation period of between sixty and sixty-three days. South of the 45th parallel, pups are born between mid-March and mid-April. In warmer regions, however, because mating times tend to be earlier, the pups are usually born between mid-February and mid-March. Throughout the wolf's range the event is celebrated with much whining and excited dancing at the den entrance by the father and the other pack members. Indeed, viewers lucky enough to witness such a scene can have no doubt that the pack is celebrating the arrival of the newcomers.

At birth, pups have short legs; blunt faces; short, thin tails; and small ears folded over their foreheads. This combination of features causes them to look more like newborn bear cubs than little wolves.

A curious captive pup, probably about eighteen days old, inspects its new world.
Peter McLeod/
First Light

Gray wolf pups huddled in their den.
Scott Leslie/
First Light

The mother has eight teats, four on each side of her stomach, and so can comfortably nurse eight pups. On average, however, she gives birth to between five and seven young. In exceptional cases, female wolves have been known to give birth to as many as fourteen pups, but at least half of such large litters would probably die from malnutrition. In any event, infant mortality rates are high, at times brought about by parasite-induced diseases, or by malnutrition resulting from sibling competition, when the strongest pups crowd out the weakest. Then too, if genetically defective pups are born — and although such births are not common, they nevertheless do occur — they will not survive.

If a pup dies in the den, the mother will likely remove it, carrying it in her mouth to some area outside and burying it, much as if she were caching a piece of meat, although in my experience, a dead pup is not later disinterred and eaten, at least not by the wolves. I have witnessed two wolf pup burials in the wild, and in recent times a Spanish film crew recorded such a burial. It followed the mother with a powerful telephoto lens as she carried the dead and already stiffened pup in her mouth to a location some distance from the den, where she set the small body aside, dug a fairly deep hole, deposited the pup in it, and then covered it thoroughly.

Pups are born blind — their eyelids are virtually glued together — and they are either entirely deaf or extremely hard of hearing, perhaps because their small ears are folded over their foreheads. Nevertheless, they emerge equipped with a sense of smell, and this allows them to find their mother's milk-filled teats just as soon as each newborn has been cleaned of the birth fluids and is nosed towards the female's stomach.

Why wolf pups emerge into their world sightless and hearing-impaired is a matter for speculation. In my view, such seemingly negative characteristics probably ensure tranquility and allow the newborns to become almost immediately and fully imprinted on the scent signature of the family, thus creating a strong and permanent bond of recognition. Then too, in the absence of sight and sound, the newborns would be better able to concentrate on feeding.

Litters of red wolves may contain four to seven pups.
© Wm. Munoz

The eyelids of the pups usually open between the tenth and thirteenth day of life. At about the same time, their ears begin to straighten and they become attuned to sounds.

The eyes of the pups are blue at first, but by the sixth or seventh week they begin to change from the outside inwards towards the dark iris, until, in the majority of cases, the sclera (the surface of the eye that surrounds the iris) eventually becomes amber-yellow. Some wolves, however, have relatively pale scleras, and others have dark chestnut scleras, the irises being a deep brown color.

The birth coat is short and finely woolly and is usually a slatey blue color, although some pups are born with a dark brown coat. At birth, pups generally weigh between twelve and sixteen ounces (340 and 450 g) and measure between ten and thirteen inches (25 and 33 cm) overall, depending on their subspecies.

Nourished on their mother's rich milk, wolf pups grow rapidly and, about one week after their eyes open, they begin to explore their nesting chamber, although, if they wander too far away at this time, the mother will usually carry them back. But they remain inquisitive, and some four weeks after birth, they will start to explore the tunnel, tottering along and falling over at times, until they reach the entrance. Once there, hesitating somewhat fearfully, they stare

European wolf cubs at two months.
© E.A. James/NHPA

myopically at the world outside while blinking repeatedly, until their eyes become accustomed to the strong light. Their initial nervousness does not last long, and they soon pluck up the courage to investigate the nearby surroundings.

At this early stage, the pups are usually watched over by their mother. In a pack, however, if the mother has gone on a hunt with other pack members, her pups are cared for by a pup-sitter. Such a guardian is usually an "aunt" or a two- or three-year-old female sibling. Wolf pup-sitters, if necessary, are capable of producing milk and are thus able to take on the role of temporary surrogate mothers in the absence of the maternal female. If the mother is killed, or if she dies from disease, the surrogate will continue to care for the young, herself being relieved by another pack female from time to time. Indeed, all wolves literally dote on the pups. Like all of their kind, the males seem to be extremely devoted fathers, and a father has been known to try to nurse his offspring following the death of their mother.

The parents of a beginning family will encounter difficulties not experienced by a well-established pack, because as the female becomes heavy with her young, she spends her time inside the den. The male must then hunt for two. Although efficient predators, lone wolves do not usually bring down prey larger than deer, so they must make do with smaller mammals. A single wolf has little difficulty keeping himself fed with such fare, but when hunting for himself as well as for his pregnant mate, he is kept busy almost full-time. After the pups are born, his nursing mate requires considerably more food in order to nourish her young properly. Later, when the pups start to eat meat, the male becomes even busier and will remain so until the pups are old enough to be left alone while both parents go hunting.

Wolf pups grow quickly. They must, for by autumn they will have to join the pack in the hunt.

Approximately eight weeks after birth, the pups weigh about fifteen pounds (7 kg), the males as a rule being about 20 percent heavier than the females. Their first set of teeth are well developed at this age, the canines being relatively long, slightly curved, and needle-sharp at the tips. Their incisors, although small, are also sharp, acting somewhat like chisels when a pup needs to cut off a piece of meat. The molars are somewhat less than half the size they will be when the permanent teeth develop at five or six months of age. Nevertheless, they are serrated and strong and are capable of crushing small bones and slicing off larger pieces of meat.

Pups begin to nibble on meat and to chew bones by the time they are about four weeks old. During their second month of life, they become weaned of their own accord and feed on the meat brought home by their parents and by other adults. The arrival of the pack is an

By ten weeks of age, the pups have been moved from the den to a rendezvous site.
© Erwin and Peggy Bauer

event that causes great excitement among the pups. Whining, with their tails wagging vigorously, the young wolves run to mob their father and other pack members, who willingly help to feed them.

Dashing up to the food bringers, the pups leap at their heads and immediately start to nip at their cheeks, behavior that stimulates the adults and causes them to regurgitate. Opening wide their great mouths, they begin to bring up undigested meat; but before the action has been completed, the first pups to reach the adults usually thrust their heads into the gaping, teeth-filled cavities and begin to eat as the food emerges from the pulsing throats. Of course, each young wolf vies with its siblings to be first at an open mouth, so the scramble to feed is fast and at times punctuated by some pup violence, although it is rare for one young wolf actually to draw blood from another.

At the same time, other members of the pack may have carried back in their mouths parts of the kill for the stay-at-home mother, or, if she has gone hunting with the pack, for the pup-sitter. In the absence of solid parts, the adults will regurgitate for the stay-at-home guardian. Meat offerings that are carried by mouth to the den usually contain bones, which are eventually left for the pups to chew.

When the young wolves are between eight and ten weeks old, the den is abandoned; it may or may not be used again the following year. The pups are moved to a location that has been termed the rendezvous site, a nursery area that is selected with great care. In most cases, a rendezvous site is chosen because it offers a number of specific features, not the least of which is some more-or-less open land that is surrounded by trees and contains a nearby source of drinking water. The open land, which is always partially covered by shrubby trees and brush, allows the pups to romp and play and to begin to practice their hunting skills by pouncing on mice or voles. The nearby forest allows the young, and indeed, the entire pack when they are all present, to hide among the trees should they be disturbed.

In such a place, watched over by all the wolves or by their mother or babysitter when the pack is out hunting, the pups romp, compete with one another for status, and copy the behavior of the adults. At times, however, they are given a quick taste of lupine discipline. The mother or pup-sitter may grab a recalcitrant little wolf by the scruff of the neck and shake it quite vigorously. As a rule, just one shake is sufficient to teach a pup how to behave.

At ten months, young wolves must share in the hunting duties with the rest of the pack. Peter McLeod/ First Light

Pups begin to demonstrate the keen intelligence of their kind almost as soon as they have left the nursery. I have personally observed a number of examples of intelligent behavior displayed by wolf pups. One such occurred when a three-month-old devised a calculated plan that was successfully put into effect. This happened in July, 1967, while I was observing through field glasses five pups and two adults in their summer rendezvous. One pup had evidently found or dug up a bone and was gnawing at it assiduously when one of its siblings approached and tried to take the bone away. The possessor planted a paw firmly on the prize and growled a full and, despite its age, ferocious warning. The intruder backed away, sat down a few paces from the chewer, and after a few minutes tried again to steal the bone, with similar results.

Four times the pup tried to take possession of the bone, every time to be rebuffed by its owner. Then, seeming to be totally disinterested in the coveted morsel, it trotted towards an area of shrubs that was about twenty feet (6 m) away from its sibling. Once there,

A yearling wolf watches an adult female as she scent-marks.
Peter McLeod/
First Light

the pup went through the motions of hunting mice or voles. It stalked, then jumped up and landed hard, front feet first, the pads pounding down the grasses and small plants. While doing this, it kept an eye on the bone chewer who, in turn, began to watch the hunter.

Presently, the pup pretended it had captured a mouse, settled on its haunches and looked as though it was about to devour the prey. This temptation was too much for the owner of the bone, who evidently decided that a tender, newly caught rodent would be better than a rather old and smelly bone. The bone-owner jumped to its feet and dashed at its sibling, who retreated as if frightened. In fact, it turned around and raced to the bone, grabbing it and running away. In vain the other pup searched for a mouse or vole. Its sibling had resorted to a clever stratagem in order to lure it away from its prize.

When the pups are about three months old, they are sometimes allowed to accompany the pack on a hunt. They bring up the rear where they are supervised by an adult. Whether these outings ever result in a kill or are just part of the tutoring of the young I cannot say. I have never seen young pups at a kill site, and I have not found in the literature any reference to the participation in a kill by the very young.

Cared for by their parents and by other pack members, the pups grow rapidly, day by day becoming stronger, bigger, and more assured as they near the time when they will go hunting with the pack, a responsibility that they will have to accept by the time they are about six months old. By then, they will have replaced their milk teeth with their adult teeth, gained some experience hunting mice and voles in the rendezvous site, developed skills and strong muscles wrestling with one another, and joined their young voices to those of the pack during frequent howling sessions.

Chapter Five

ORPHAN WOLF PUPS

On the morning of May 20, 1984, my wife Sharon and I were both worried as we drove away from Whitehorse, the capital of the Yukon Territory, on our way home to Ontario. The causes for our concern were on the back seat in a cardboard box — two twenty-three-day-old wolf puppies. They were in such poor condition that we did not really expect them to survive the 3,400-mile (5500-km) journey to our property.

One pup, the male, was black, except for a small white star on his chest. If he survived, there was no question that his adult coat would be black. The coat of the other pup, a female, was dressed in a mix of fur that was dark gray, white, and tan. The cubs were about the same size, little bundles of soft fur that weighed somewhat less than one pound (0.5 kg) apiece at a time when they should have scaled at least twice that amount. Their deplorable condition was due to the fact that for the last twelve days they had been fed full-strength cow's milk four times a day, instead of a balanced infant formula every four hours, the diet and timing that is required in order to raise healthy pups when they are being nurtured by human foster parents.

As a result of their improper care, the pups were suffering from severe diarrhea, their little bodies were wasted, and they were

Gray wolf pups gather outside their den in Alaska.
© Art Wolfe

seriously dehydrated. They were the last survivors of a litter of six, and evidently the strongest; otherwise, like their siblings, they too would have died from malnutrition and colitis days before they came into our care.

We had accepted the responsibility for the pups the evening before, but had been unable to obtain formula because the drugstores were closed. Fortunately, the young man who had tried to raise them had given us a nursing bottle, although the only nourishment we had been able to obtain from the hotel was skimmed milk. This had to do until we could buy the right kind of formula. We gave it to them through the night, Sharon and I taking turns feeding the pathetic little creatures every three hours, instead of every four. We reasoned that they needed more frequent bottling because of their condition and of the unsuitable diet, most of which appeared to be excreted almost as quickly as it was ingested.

We had arrived in Whitehorse the previous morning and, after we checked into the hotel, we had gone downstairs to have lunch. As we were waiting for the first course, we discussed wolves. We each commented on the stupidity of governments that, for the sake of political expediency, continually bowed to lobby groups whose members claimed that wolves slaughtered animals that they, the "sportsmen," wanted to kill. To add to the fables, they continued to insist that wolves attacked humans.

After a while, I noticed that a young man sitting at the next table seemed to be listening to our conversation, and I wondered if he would tell us what he thought about wolves. Before I could think of a way to approach him, he got up and walked to our table.

"Excuse me," he said. "I couldn't help hearing what you were

A female wolf follows her mate along a winter trail.
© Thomas Kitchin

talking about. Fact is, I have a bitch wolf. Had her for three years now. Elsa's her name. Well, I had to keep her tied up and she got mated by a wild wolf. I rent this cabin outside of Carcross."

I invited him to sit down and he introduced himself as Pete King. He was a diamond-driller, and he usually travelled in a camper truck with his wolf. In December, Pete had been laid off work, so he rented the cabin. He explained that while he was in the field at drilling sites, he kept Elsa with him. He walked her a lot, and she almost always slept with him in the trailer. But because he was now looking for work, he had been forced to keep her during the day on a long chain that was attached to the cabin while he travelled to nearby areas that might require an experienced driller. He had not realized when he started out in search of employment in mid-February that the wolf mating season had begun in southern Yukon. Elsa, redolent with mating odors, had evidently attracted a lone male wolf, but knowledge of her mating was not apparent to Pete's inexperienced eyes until Elsa had delivered herself of six pups.

I was not unduly surprised by the fact that the captive bitch had been mated by a wild wolf. In the northlands, especially where urbanization has taken place, wolves at times prowl around settlements, on occasions killing dogs that are allowed to run free. Now and then, if a lone male wolf is prowling in search of a female, he may mate with a husky bitch that has entered the estrus (heat) period. But I found it strange that Pete had not in due course become aware that his wolf was pregnant.

"Why didn't you realize early on that Elsa was pregnant?" I asked him, wondering if he was telling me the truth.

"I just didn't know. I saw she was getting fat, alright, but I

An Alaskan wolf, with an arctic ground squirrel in its mouth, returns to the den to feed its pups.
© Ron Sanford

figured it was because she wasn't getting enough exercise. It was sure a surprise when I woke up in the cabin one morning and saw her lying on her side, feeding six pups!" he said.

Later, while in Whitehorse, he began to offer free wolf puppies to whoever would take them, for, as he explained, he could travel with Elsa alone, but what could he do with seven wolves? But nobody would accept the little whelps.

Eight days later, while Pete was away from home, somebody who evidently hated wolves had driven to his cabin and, seeing Elsa lying stretched out on her side, had used a shotgun to pepper her with bird-shot. The wolf was not severely wounded, but most of the pellets had hit her stomach and injured her mammary glands. She could no longer feed the pups. Two weeks had gone by and Pete had tried to feed the little wolves, but without success.

"Now, only two are alive. Will you take them?" he asked.

Having some years earlier acted as foster parent to two Ontario wolf pups, I knew the many responsibilities involved in such a task. Staring into space, I shook my head. Then I looked at Sharon. She was openly glaring at me.

"What do you think?" I asked her.

"I think we should go and look at the surviving pups. And I think that Elsa should be seen by a vet," she replied firmly.

I agreed. We drove to Pete's cabin in Carcross and, of course, we said we would take the pups.

When we picked up the pups that afternoon, we arranged with Pete to have Elsa properly cared for by a veterinarian. As we were to learn later in a letter scribbled by the young man, the wolf recovered. While surgery was being performed, the veterinarian did an ovario-hysterectomy to ensure that Elsa would never again become pregnant.

After our sleepless night with the pups, Sharon and I were determined to save them if possible. At 8:45 the next morning, while Sharon was trying to console the starving little animals, I left the hotel, walked to a nearby drugstore, and waited for its doors to open. Half an hour later I had bought another feeding bottle, several rubber nipples, a can of powdered infant formula, and a carton of infant cereal. Having already had considerable experience in the raising of young mammals, I had also bought two rolls of paper towels, two toilet paper rolls, a pot of honey, a container of glucose, a quart-capacity vacuum flask, and a package of disposable diapers, which I would have normally shunned at home in favor of the washable cloth kind, but which on this occasion were necessary in order to keep clean the cardboard box which would serve the pups as a den while we travelled.

On my return to the hotel, we mixed the formula with slightly more water than was called for on the instruction label so as to ensure that it would not be too rich for the pups. We added some rice Pablum in hopes of tightening their loose bowels, a little honey for taste, and some glucose. The latter was added to ensure that the absence of glucose in their previous whole-milk diet would not lead to the formation of cataracts in later life. This condition often occurs when wolf and dog puppies are removed from their mothers before the second week of life and then fed a glucose-deficient diet.

By 9:45, we had warmed the formula to blood temperature and were prepared to feed our yowling little wards, selecting the female first because she had pushed ahead of her brother as soon as she

This five-week-old pup's blue eyes will soon begin to change to amber–yellow.
© Erwin and Peggy Bauer

smelled the food. We had mixed eight ounces (230 mL) of the preparation, but I intended to allow each of them to take only about three ounces (100 mL) on this, their first proper feeding since they had been deprived of their mother's milk.

Sharon, sitting cross-legged on the floor, her back against the hotel's bed, picked up the tiny pup and offered her the bottle. She took one sniff, fixed her large mouth on the nipple, and began to suck lustily. Three ounces were quickly sucked out of the nurser; but when Sharon tried to take the nipple out of the pup's mouth, she clung to it with surprising strength, grasping the bottle with both front paws while standing as tall as she could on her back legs, trying to follow the disappearing nipple.

"The poor little thing is *starving*," lamented Sharon. "Let her have another ounce."

Almost before my wife had finished speaking, the pup had pulled her hand down and fastened her mouth anew on the nipple. That other ounce was soon sucked down! Now, against her struggles and vocal protestations, the bottle was forcibly removed and I picked her up, a paper towel in hand, and began to wipe her mouth and face while Sharon introduced the bottle to the male.

He immediately grabbed the nipple and began to suck. Although not quite as rambunctious as his sister, he also refused to let go of the teat after ingesting three ounces (100 mL) and was allowed to continue to suck until the bottle was almost empty; then, rather like his sister had done, he grasped the nurser with his front paws and refused to let go until all that was left were a few bubbles. Both pups now began to whine, obviously wanting more food, but I was firm in my refusal. It would have been dangerous to allow them to overfeed, for starving animals that are given food will continue to eat long past the point of satiety, subconsciously prompted to do so on the premise that they might not find anything more to eat for a long time.

While Sharon was wiping the male's mouth and face, I dipped a face-cloth in warm water and massaged the female's abdomen and rectal area to stimulate her bowels, as the mother would have done with her warm, wet tongue. Moments later she defecated, her movement runny and evil smelling. After she was cleaned up, we swapped pups. Sharon now held the female and I massaged the male. The black pup performed much as his sister had done, and the result was equally odorous.

Soon afterward, I removed an old wool cardigan from my suitcase and used it to make a warm bed in the cardboard box. After covering the sweater with a spread-out disposable diaper, I placed the suddenly sleepy twins inside the container and closed the four lid sections by overlapping them so that there were enough open spaces for ventilation.

Gathering up our things, I paid the hotel bill and packed the car. When Sharon came outside carrying the box of pups, we placed it on the back seat and began our return journey.

Returning Home

Musing about our unexpected responsibilities, I felt torn between anxiety to get home so that we could devote ourselves to the care of the little waifs and my disappointment at having to forgo our trip to Alaska, a journey planned after we had spent two weeks studying a captive pack of wolves in Michigan's Upper Peninsula.

A wolf crushes a bone with its efficient carnassial (molar) teeth and its powerful jaws.
© J.D. Taylor

When we arrived home from Michigan, I had found a series of reports from various Canadian and American conservation groups among the pile of letters and circulars that were waiting for us. The reports detailed the persecution of wolves by provincial and state governments in Alberta, British Columbia, the Yukon Territory, and Alaska. The wolf-kill programs in these regions were in response to the spurious claim that wolves were decimating prey species. In most cases, the animals were being shot from helicopters and low-flying fixed-wing aircraft. In some instances, they were being poisoned with the abominable compound 1080 (sodium fluoroacetate). This virulent venom does not break down in the bait or in the dead body of its victims and thus can kill through chain-reaction a variety of other mammals and birds. Victims of 1080 suffer an agonizing death.

The news was shocking. By then, I had been studying wolves for almost thirty years. During that time, with the help of my late wife Joan, I had already raised to adulthood two orphan wolf pups, which we had eventually been able to reintroduce into their own habitat. I had also observed at length a number of wild packs. Over the years, I had come to respect wolves for their intelligence and for their marvellous family relationships. For those reasons, and, I must admit, because I like wolves, I felt that I should personally visit the northwestern kill regions to learn more about the slaughter. In addition, I intended to conduct in the areas involved a survey of public attitudes towards the wolves and other predators. Sharon, as usual, volunteered to come with me to keep me company and to record our findings.

For the first four hours of our return trip, I drove in silence, thinking that it was somewhat ironic that apart from getting first-hand information about the Alberta and British Columbia wolf-kill projects, our hopes for obtaining more data in the Yukon and Alaska had been swiftly dashed by the needs of two tiny wolves. After a time, however, I realized that, although there was nothing much that Sharon and I could have done to immediately cause the governments involved to stop slaughtering wolves, we had been given an opportunity to try to save the lives of the two pups. This was now our goal, and, if we succeeded, we would be able to take considerable satisfaction from the knowledge that at least two Yukon wolves had survived. Later, I thought, through my writings and by joining with other like-minded people, we could perhaps cause the wolf killers to stop the slaughter.

I felt then, as I do now, that it is unconscionable to kill predators simply because so-called sportsmen claim, without data to support their accusations, that the wild hunters are reducing the numbers of deer, elk, moose, or other so-called "game species" in a particular region. Such behavior is a symptom of the social ills that have caused our species to make war against its own kind and, in fact, to make war on all living things that do not fit into the "useful for humans" category. It completely disregards the fact that in nature everything is connected to everything else; if one part of an ecosystem is removed, the entire system suffers.

This type of thinking had caused me to undertake our journey west. I was especially outraged when I read in *Nature Canada*, the magazine of the Canadian Nature Federation, that the wolf slaughter in British Columbia was being partly financed by a hunter group, a local guide-outfitter in the region, and American wild-sheep killers.

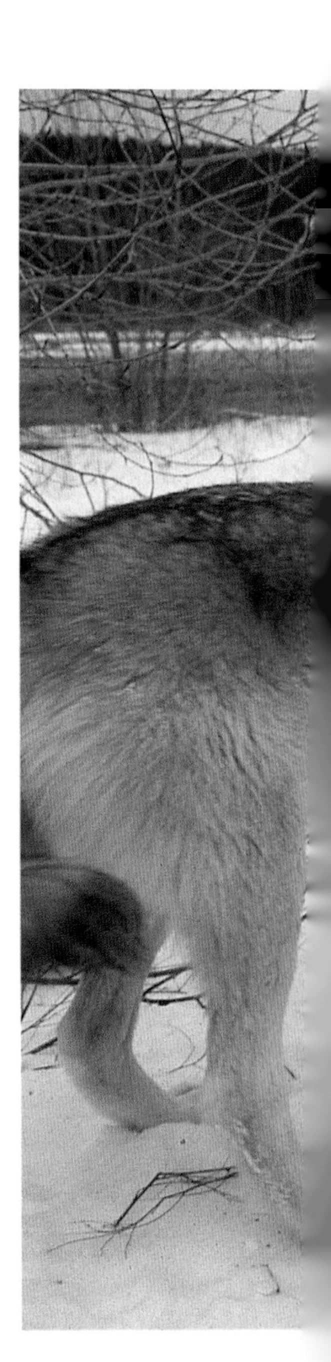

Wolves feed on the carcass of a bull moose.
© Erwin and Peggy Bauer

"We were amazed to learn of a raffle to raise funds for a massive wolf kill north of Fort St. John, British Columbia," reported the magazine. "The B.C. Wildlife Federation and the Northern B.C. Guides Association have sponsored a raffle to pay some $100,000 to the B.C. Environment Department to shoot as many as 400 wolves from helicopters. The intent is to increase numbers of elk, moose, and caribou."

In addition, the Foundation for North American Wild Sheep, whose members helped to kill almost 1,500 sheep in the United States in 1980, donated $100,000 to help the British Columbia government's wolf-kill fund. In all, the three sponsoring groups contributed a minimum of $200,000 to the ill-conceived plan.

Meanwhile, the Wildlife Biologists' Section of the Canadian Society of Biologists, having studied the documents that were claimed by the B.C. government to show justification for the killing of wolves, also condemned the slaughter. In a position statement to the government, the biologists noted that there was no biological basis "or biological justification for the wolf control programs currently being conducted in northeastern British Columbia."

Twenty-four hours after we left Whitehorse, we reached Dawson Creek, British Columbia. From there, 2,500 miles (4000 km) still separated us from home, but we were heartened by the condition of the little wolves. They had improved dramatically.

Late in the afternoon of May 21, we reached Edmonton, Alberta, after having driven 352 miles (566 km) from Dawson Creek. After driving through the busy part of the city, we stopped at a wayside restaurant to have an early dinner and to get hot water for

A mother wolf with three pups. Two of them are wrestling beside her.
© E.A. James/NHPA

a new batch of formula. By the time we had eaten, four hours had elapsed since the pups had been given their last feed, a fact that was announced by small howls as we opened the back door of the car and placed their box between us on the front seat.

As soon as the lids were opened, the female stood upright, whined, and scrambled out of the box, her movements unsteady, but very determined. The male followed suit, not as agilely as his sister, but equally determined. Sharon was reaching to put the female in her lap, when the little wolf climbed there of her own accord. The male, meanwhile, had made himself at home on my lap.

Each pup devoured four ounces (115 mL) of a formula mix in which we had replaced the rice Pablum with mixed cereal Pablum and had added about two ounces (57 mL) of "chicken puree" baby food. After they had eaten and had been cleaned up, I examined what they had left on the disposable diaper; although the stools were soft because of the diet, there was no diarrhea and no foul odor. We now became so confident that the pups would survive, that we named them. We called the female Taiga (pronounced *Ty-gah*), a term used to describe the boreal forests of North America and Eurasia, and we named the male Tundra, which describes the relatively treeless plains of the circumpolar boreal regions.

We arrived home during the evening of May 25, settled the pups in their nursery box, had a late supper, and then went to bed ourselves after setting the alarm clock for midnight, when we would feed our wards. That became our routine for the next two weeks. Feeding times were 4:00 a.m., 8:00 a.m., noon, 4:00 p.m., 8:00 p.m., and midnight.

The next morning, after their eight o'clock feeding, we measured

Taiga, the eight-year-old female wolf that was rescued by the author and his wife when she was twenty-three days old.
© Bob Gurr Photo

and weighed the pups. Both measured exactly seventeen inches (43 cm) from nose tip to tail end, and both had evidently gained some weight, although because we had been unable to weigh them at the start of our journey, we could not determine how much. When we got them, their bones could be easily felt through their coats, and now their bones were not as prominent. At twenty-nine days old, they should have approximated the weight of northern wolf pups of that age, which would be about 4½ pounds (2 kg) for the male and slightly less for the female.

It is not an easy matter to weigh a wriggling wolf puppy. Tundra, after giving us considerable difficulty, scaled three pounds twelve ounces (1.7 kg) and Taiga, after presenting even more difficulties — one of which included urinating all over the scale — weighed in at three pounds six ounces (1.5 kg). Considering the condition in which they had come to us, I was more than satisfied with their progress and somewhat surprised at the speed of their gains.

For the next seven days, Tundra and Taiga shared our bedroom during the night and spent part of the daylight hours in a six-foot by three-foot (2-m by 1-m) nursery that I had made for them and kept in my office. The pen, with two-foot-high (60-cm) sides, was "paved" with artificial turf and contained a proper nesting box situated in one corner. A spare section of the "turf" was kept handy, so that we could exchange the soiled one for a clean one when necessary.

On the eighth day, we cut back on the feeding time and increased the quantity of formula to six ounces (170 mL). We gave them a last meal at midnight, cleaned them up, and then took them downstairs to their pen, where they spent the night. We meant to give them their first feeding of the day at eight o'clock, but on that occasion, tired after so many disturbed nights, we forgot to set the alarm for seven o'clock. We overslept, only to be awakened suddenly at 8:30 by the constant, loud, tenor howls of two wolf puppies from downstairs.

I had started keeping a record of Tundra's and Taiga's behavior as we travelled, and I continued doing so when we reached home. Looking now at the record, I note that the day after we had introduced the pups to the office pen, I settled them in it soon after lunch. As I sat down to write, I noticed that Tundra had evidently gone to sleep inside the den box, but that Taiga was sleeping on her back in one corner of the pen.

I wondered if the noise of the typewriter would disturb them. Watching Taiga, I hit a few keys. She remained fast asleep and Tundra, although unseen, told me by his silence that the clacking of the keys had not disturbed him either. In fact, I did not find their behavior unusual, inasmuch as both pups had quickly become accustomed to the household sounds, even to the vacuum cleaner's racket.

I soon became engrossed in my work and remained so for about an hour, when I was in need of a reference book from the shelf above my desk. Before reaching for it, I looked at Taiga; she was still asleep in the same position. I reached up to slide the book out of its place, continuing to watch the pup. It made a slight scraping sound, one that was foreign to the wolves, and even before I had finished taking it off the shelf, Taiga, who was still tottery when walking and stumbled a good deal when she tried to run, now leaped to her feet in one swift motion and streaked into the nesting box without a stumble.

When an animal is startled, its endocrine system instantly floods the bloodstream with a variety of hormones, especially adrenalin, the hormone that triggers the so-called "flight or fight" response in all mammals. Taiga had just given me an elegant demonstration of the power of that hormone, and there was more to come. When I called her and she re-emerged from the den, she was again tottery. The dash to shelter had used up the excess hormones, and afterward she had immediately returned to a metabolic condition that was normal for her age.

After that entry, for the next week or so, my daily records were fairly repetitious, noting that although the pups were growing rapidly and eating well, their development was not particularly rapid. But by June 3, this situation began to change, as the following excerpts show:

By the time the pups are ten to twelve weeks old, they have usually determined their status with one another.
Peter McLeod/ First Light

> June 3: Fed raw chicken again this a.m. Both eat well, masticating the food very effectively before swallowing. Later this day both became disturbed for reasons unknown. They whined a good deal, howled, and were generally restless. Yet they had earlier fed well and they were not now hungry, refusing food when it was offered. We petted them gently, stroking, talking to them softly. They started to run about, chasing each other. Soon afterwards they were put in the pen and they went to sleep. Were they over-tired?
>
> June 3, 4:00 p.m.: Pups had a good meal, including raw beef and a bone each. After food, they began to play-fight, during which both held their tails high, each seeking to be dominant and refusing to submit to the other.
>
> June 9: Pups have weaned themselves. At feed time this a.m., they refused formula, but gobbled up raw chicken and beef hamburger;

> Tundra ate half a pound [0.2 kg]; Taiga, always the bigger eater, ate eleven ounces [0.3 kg]. This p.m. (3:30) Taiga moved bowels on the living room rug, turned around, sniffed at the firm stool, then tried to scratch-kick around it, displaying territorial marking for the first time. Sharon was angry at first, but she quickly relented when Taiga, noting her distress, ran away and hid under the sofa, followed by Tundra. I had to lie flat on the floor and coax them out. As with the first two pups that I raised [Matta, female; Wa, male], the pups can immediately detect distress in both of us, be that anger, sorrow, or just plain worry.

Growing Up

When Tundra and Taiga first came into our care, even as I was driving home from the Yukon Territory, I decided that I would eventually rehabilitate them and return them to their own world, as I had done with Matta and Wa. By the time the pups were three months old, however, I realized that such a release was beyond my financial means.

Times had changed radically since Matta and Wa were released. By the 1980s, development in Ontario had spread east, west, and north. What had been small, relatively isolated communities in the mid-1960s were now thriving towns, and lakesides had been developed for recreational lodges and cottages. Wolves, on the other hand, were still persecuted with as much hatred as ever.

In order to find a suitable habitat in which to release Tundra and Taiga, it would have been necessary to charter an aircraft (at a rate of over $150 an hour) and fly over vast regions of northern Ontario, looking for a range that was far enough away from civilization to

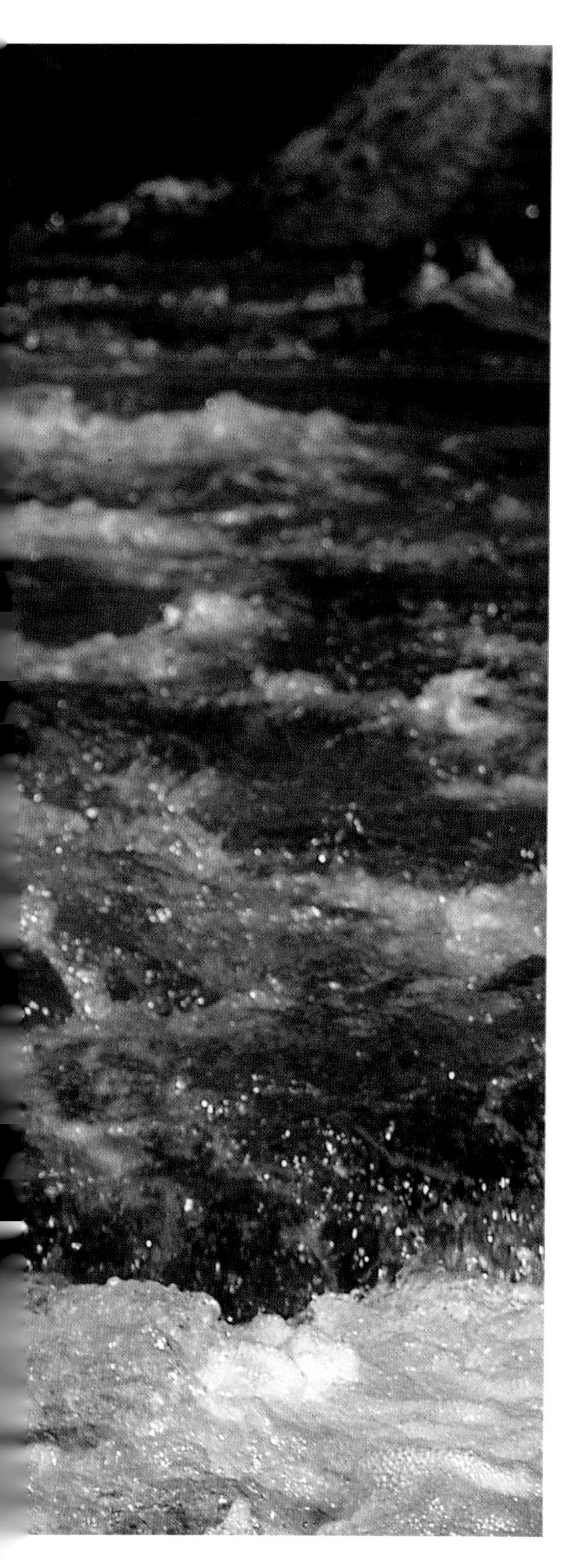

This pup is exploring near a rendezvous site. From six to ten weeks of age, a pup's coordination increases rapidly.
© Erwin and Peggy Bauer

ensure the following essentials: that the young wolves would not be hunted down; that there would be sufficient natural prey available to them; that there would not be a resident wolf pack to compete against them; and, finally, that the region would offer me a location where I could haul in building materials to construct a makeshift cabin in which to stay while Tundra and Taiga learned the ways of the wilderness. That last item alone presented serious difficulties for me because I had no idea how long I would have to stay in the wilderness in order to act as their trainer and protector. When I calculated what such a plan would cost, and arrived at $30,000 (which I did not have), I decided to build our wards a large, natural enclosure at home.

Assisted by Sharon and Murray Palmer, a young biologist friend who had at times assisted me during field studies, we first constructed a temporary enclosure that measured 175 feet by 60 feet (53 by 18 m). By early September, we enlarged this enclosure to incorporate 1.6 acres (6500 m^2), which included forest, some open area, a barn where they could shelter during the height of the fly season, and a pond that supplied them with constant clean water. In this habitat, our wards had turned into handsome young wolves by mid-September. Tundra weighed forty-nine pounds (22 kg) and Taiga was just three pounds (1.4 kg) lighter.

Earlier, when they were six weeks old and before we had placed them in the small enclosure, we had got them accustomed to wearing collars. Once they were comfortable wearing those devices, we began to take them for walks within our 100-acre (40-ha) forest property, each secured by an extension lead. The collars, however, were worn only when we took them on such outings.

Wild wolf pups at that age already begin to exercise hunting skills, at times thumping down hard with their forefeet to try to stun mice or voles that are scurrying beneath a carpet of wild grasses and sedges; and on other occasions chasing squirrels or even leaping to try to catch birds in flight. I had watched such behavior in the wild on many occasions, but I had not been able to determine whether it was prompted by genetic imprinting or by a desire to copy adult behavior. I leaned towards genetic imprinting, but I could not be sure.

Tundra and Taiga quickly proved that the genetic imprinting theory was valid. The first day that we took them for a walk, they led us to a swale that exists just behind our dwelling and there, leaping and thumping with their paws, as agile and determined as their wild kin had been, they hunted mice. They missed quite a few, it is true, but they did catch some. Then, a week later, while we were walking through a thick stand of young larches, Tundra and I in the lead, the pup suddenly darted forward, leaped upward, and caught a sparrow in flight.

In mid-September, while we were taking the pups for a walk, Tundra stopped suddenly and raised his head, his nostrils flaring as he caught an interesting scent. A moment later, he turned and led me towards a small pond. Inasmuch as Wa, at about the same age, had behaved similarly and had led me in a beeline to a porcupine located more than a mile away, I gave Tundra his head, wondering what he had scented and how far he would pull me in order to locate its source.

He trotted along eagerly for 350 yards (325 m) or so, which brought us to the edge of the pond. Once there, he circled to the back side of the little lake and stopped, nose down and within one inch (2.5 cm) of the track left by an adult wild wolf, which had obviously

A healthy white-tailed deer stag can usually outdistance a wolf. In general, wolves prey on weaker individuals.
© Thomas Kitchin

visited the pups sometime during the previous night. The track was fresh and had been set down on a small patch of mud on the shore. Tundra, after sniffing the spoor, then stepped forward and left his own, much smaller imprint just in front of the one left by his relative.

Ten days later, while walking in the forest, I saw a deer some distance to our right. Tundra, much lower to the ground, did not actually see it, but he immediately picked up the animal's scent in the tracks that it had left as it had trotted across our line of travel. The young wolf, however, began by following the scent backwards, away from the target. But not for long! After advancing about six or seven yards (5 or 6 m), he quite clearly realized that the scent was weakening, and he immediately swung around and began to track in the right direction. The deer, of course, had by now glided away, so, after Tundra led us to a small pile of fresh deer manure, deposited out of nervousness as the animal left the scene, I persuaded the now recalcitrant little wolf to turn in a different direction.

These and other demonstrations of the acuity of Tundra and Taiga, as well as similar examples witnessed in the wild, were to give me a much deeper understanding of the social order of wolves, of their behavior, and of their magnificent intellect.

Wolf Intelligence

There is no doubt that the wolf is a highly intelligent mammal, which is capable of learning from experience and is able to think in order to solve problems that it has never before encountered. Somewhat like the wild pup that had fooled its sibling into giving up its prized bone, Taiga demonstrated her intelligence — her ability for

Tundra, Taiga's brother.
© Bob Gurr Photo

scheming — one afternoon six years ago while I was adding an overhang to the enclosure.

I was atop a ladder outside the enclosure and a helper was assisting from the ground. Sharon was standing some distance away, watching the operation, which involved a one-hundred-foot-long (30-m) rope that I was using to stretch the fence wire. I did not, of course, need such a long rope for the job at hand, but neither did I want to cut it. The result was that most of the rope was lying on the grass outside the enclosure at a distance that I had concluded was too far for either of the two wolves to reach.

In any event, as I was about to hammer a staple in place, I was almost knocked off the ladder when the rope suddenly disappeared into the enclosure, carried away by Taiga. I had not seen the wolf capture the rope, but Sharon had observed the whole affair.

Taiga had first tried to reach the rope from a standing position by sticking her paw through one of the spaces in the mesh. When this failed, she dropped on her haunches, belly down, and tried to stick her muzzle through the mesh aperture. She failed again. Then she lay stretched out on her side and thrust her muzzle sideways through the wire, only to fail once more. But now, instead of retracting her muzzle, she stuck out her long, elastic tongue, which was able to reach the rope. Next she managed to thrust her tongue just under the end of the rope and began to flip the rope towards her, continuing until it was close enough to her mouth for her to grab it, pull it into the enclosure, and then race away at high speed.

This was the first unquestionable demonstration of Taiga's intelligence. And Tundra, although not as likely to tease us, demonstrated on many occasions that he, too, could think very effectively.

Chapter Six

WOLVES AND HUMANS

The greek biographer plutarch (A.D. ca. 46–120), in a collection of his writings, *Parallel Lives*, noted the payment of what may have been the first documented wolf bounty when he wrote that a reward of five silver drachmas (the principal coins of ancient Greece) was paid by Greek officials to a hunter who had brought in a dead male wolf. Some time before that, Pliny the Elder (A.D. 23–79), the respected Roman naturalist, alluded in his writings to the conflicts that then existed between humans and animals, especially wolves. Later, in the Statutes of Charlemagne (A.D. 742–814), it is recorded that two hunters were to be employed in each French community to destroy wolves.

From such historical details, it is evident that the war against wolves, as well as the payment of bounties to those who kill them, began not less than 2,000 years ago, and probably originated with the advent of farming about 8000 B.C. This revolutionary event put the human species on a collision course with predatory animals and with the natural habitats that predators and their prey species require in order to survive. Circumstantial evidence supports the hypothesis that long before the emergence of Greek and Roman civilizations, farming, especially the raising of domestic livestock, was responsible

Two European wolves. By the early part of this century, wolves had virtually disappeared in most of Western Europe.
© Jean-Paul Ferrero/ Auscape

for creating the first confrontations between humans and wolves. Later, as human populations increased, more and more land was cleared for villages and towns and for the growing of crops and the rearing of food animals, resulting in the shrinking of wolf territory as well as the territory of their prey. Wolves, deprived of sufficient wild food, began to feed on domestic stock as well. At the same time, the prey species, whose food base was also reduced, began to supplement their natural foods by feeding on agricultural crops which, in any event, were more palatable and easier to obtain. Such occurrences undoubtedly caused men to kill more browsing animals, such as deer, to protect their crops, as well as for human consumption. Much later, although no longer having to rely heavily on wild animals for food, men continued to hunt, but for sport.

As its forest habitat continued to shrink and the populations of prey species fell drastically, the wolf had no option but to prey heavily on domestic stock. In doing so, it exacerbated man's hatred, and folk mythology began to portray the wolf as a demon, an evil being put on earth to torment humans.

Some 900 years after Plutarch recorded the five drachma bounty, Norse mythology added to the unmerited evil reputation of the wolf. A satanic overlord, Loki, was introduced in the *Edda*. Loki was seen as a being who contrived all manner of frauds and mischief. He was supposedly handsome and of a giant race. The first of his three children was the enormous wolf, Fenris, who devoured many people before the gods were able to chain him.

Similarly, another ancient myth re-emerged in the Middle Ages. It was believed that certain evilly disposed men had the power to transform themselves into werewolves. These individuals went out at

"Both werewolves and man-bears are well known from medieval literature. The so-called berserks often formed the bodyguard of early Scandinavian kings and chieftains."

FROM *SCANDINAVIAN FOLK BELIEF AND LEGEND*, REIMUND KVIDELAND AND HENNING K. SEHMSDORF, EDITORS

night to kill helpless people, and although they did not eat the flesh of their victims, they did drink their blood. The werewolf myth soon flourished in France and in other countries of Europe and is still believed by villagers in remote regions of central Europe.

In fact, it appears that the so-called werewolves were the victims of an inherited disease called porphyria, a metabolic disease that is today amenable to treatment. Porphyrins are a class of red-pigmented compounds that form the active nucleus of chlorophylls and hemoglobin. Porphyria sufferers excrete large quantities of porphyrin in the urine. Because the compound is red-pigmented, sufferers evidently believed that they were passing blood. There is also evidence to suggest that some individuals suffered from lycanthropy, a belief that they were wolves, a condition that is diagnosed occasionally even today among schizophrenics.

It is interesting to note that Pliny, who was an astute scientist in his day, was aware that some people believed that men could, indeed, turn themselves into wolves. He wrote, "That men can change into wolves and then back to men once more I shall dismiss absolutely as nonsense. Else we would have to accept all the stories that our extensive research have proved to be false." Pliny's terse comment clearly established that the werewolf myth arose either before the birth of Christ or very soon afterward.

In addition to the werewolf myths, there is a legion of other patently ridiculous stories, including such childish tales as "Little Red Riding Hood" and "The Three Little Pigs." Some examples portray the wolf as a gluttonous, blood-thirsty, sadistic creature, which kills and eats humans, and invariably kills more animals than it can eat at one meal. One tale says that a wolf deliberately attacks a

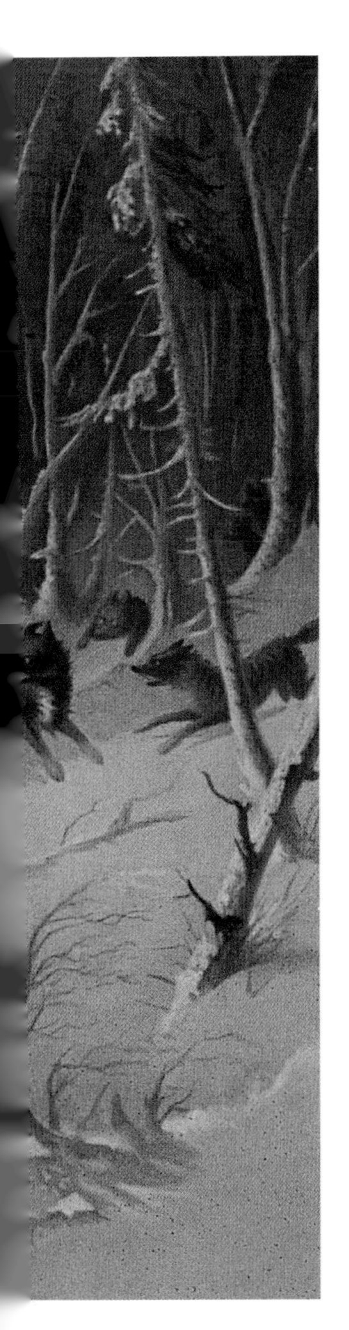

In this fanciful late nineteenth-century print, the artist, William Raphael, depicts French-Canadian habitants being chased by wolves. National Archives of Canada C-22095

pregnant deer, slashing open her stomach to rip out her fetus, which it then eats while the mother is allowed to watch the entire procedure before she herself is killed.

I have heard such stories everywhere I go. Invariably, the teller claims that he has either experienced such events at first hand, or his father has, or an uncle, or neighbor has experienced them. Of course, there is never any proof offered.

In the United States, there is great opposition to the federal government's plans to reintroduce the gray wolf into Yellowstone National Park, which occupies territory in the states of Montana, Wyoming, and Idaho, in the middle of ranch country. Despite the fact that the wolf was designated an endangered species in 1973 — which under the law should have begun its reintroduction in suitable habitats across the United States — opposition from ranchers has been so strong that only in 1991 did Congress direct the United States Fish and Wildlife Service to produce an environmental impact statement, which is the first step to be taken when reintroducing a species into its original habitat. The study is not expected to be completed until 1994, when it will be studied by the Secretary of the Interior, who will then decide whether wolves will be reintroduced into Yellowstone, where they were exterminated more than forty years ago.

The hatred of wolves appears deeply embedded in the human psyche, as becomes evident when one reads John Pollard's *Wolves and Werewolves.* In that book, the author recounts horror-story after horror-story of wolves killing humans, threatening them, and raiding farmsteads. In one literary flight of fancy, he reports that a dog, escaping a wolf, had to seek shelter on top of a haystack. I found that one amusing, because if a dog could climb up there, the wolf would be able to do so as well, probably arriving ahead of its supposed prey. The book contains a litany of veritable nonsense of which I quote but one example.

In one story, purported to have occurred on January 26, 1914, in France, a little girl was supposed to have been killed by a wolf as she walked home through the woods (shades of Little Red Riding Hood). The author writes that she had a "queer feeling that she was being watched . . ." and this made her quicken her step. Inasmuch as the tale notes that the child was killed that evening and that her remains were found in the forest the next day, one is left to wonder how the author learned about the girl's feelings as she walked through the woods. But then, myths do not have to stick to the unvarnished truth. Unfortunately, some people seem to enjoy that sort of stuff, and even more unfortunately, truth can be killed more easily than fiction.

In this sixteenth-century illustration, the wolf is used as a symbol for the devil while the sheep represents the Christian church. Courtesy of University of Chicago Library, Department of Special Collections

Wolf Attacks

There is no doubt that wolves, or wolf hybrids, did attack humans in Europe during historic times. It is equally true to state that in North America wolves did not, and do not, attack humans, although many people claim that such attacks have taken place. In fact, there are only two documented cases of so-called wolf attacks occurring in the New World. Both took place in Canada.

The first of these involved a scientist employed by the government of Canada. The incident occurred in 1927 in the Northwest Territories where the scientist and two companions were doing research. They were camped on the tundra, their dogs tied to stakes.

The story of Little Red Riding Hood perpetuates a number of myths about wolves.

Early one morning the huskies began to bark. The men emerged from their tent and saw a wolf that had come into the encampment; it was a female and almost certainly a loner that had left her pack to seek a mate. She had evidently been attracted by the odor of the male dogs, which, unlike male wolves, are always sexually active. The men immediately began to stone the wolf, and the scientist in question picked up a large rock and ran at her, raising the stone above his head ready to bring it down on the intruder. She dodged, and as his arms came down near her, she bit one of them and ran away. The man realized that the wolf had, understandably, sought to defend herself. He did not report the injury, and his wounds healed cleanly. More recently, the incident was classed as a wolf attack!

The second confrontation occurred in the late winter of 1942 in the Chapleau region of Ontario. A railroad section foreman was driving his speeder cart along the rail line and was knocked off the machine by a charging wolf. The man was able to get up out of the snow and to grab two axes that were on the speeder's platform while the wolf was still some fifty feet (15 m) away. Moments later, the wolf charged again, but the man fought him off with the axes, losing one as he sought to hit the animal. The man reported that as the wolf charged, it was growling and gnashing its teeth. The fight, according to the man, lasted about forty minutes and was brought to an end when a train arrived and its crew jumped off, armed with tools, and killed the wolf. The man was not bitten. The account of the attack was documented in the *Journal of Mammalogy*, published in 1947 by the American Society of Mammalogists. The wolf's behavior strongly suggests that it was rabid, but rabies tests on dead animals were not common at that time.

In contrast, the late Dr. Adolph Murie, an eminent biologist employed by the United States Fish and Wildlife Service, had many close encounters with wolves in Alaska, but was never attacked. In at least one situation, he offered what would be considered severe provocation to a wolf family. He crawled into a den, removed three very young pups from a litter of six, crawled out with them in the presence of both parents, studied them, put one in his packsack, returned the other two to the nesting chamber, crawled back out and left the area without having been subject to more than some barking and howling.

Why, then, did wolves attack people in Europe but not in North America, where they continue to avoid doing so today? The probable explanation may be that during the Middle Ages or perhaps earlier, Europeans developed breeds of large, mastiff-like dogs. They were intended to keep wolves away from farms, but a great many of these dogs were allowed to run free, and old records indicate that they mated with wolves to produce large hybrids. These wolf-dogs were far more powerful than the somewhat small European wolf, which weighs about sixty-five pounds (30 kg). The hybrids were less fearful of humans, and they were more aggressive, a fact that is documented in the story of "The Beast of Gévaudan." This story was compiled from old French accounts and published in 1901 by Abbé François Fabre. The facts of the case were taken from parish records and other documents.

The "wolf" attacks occurred in the regions of Gévaudan and Vivarais in France between 1764 and 1767, and, although the narrative refers to "beast" (singular), it evidently involved two animals. During almost three years, the records show that over a hundred

An early sixteenth-century European drawing of a wolf or werewolf attacking two men.
Courtesy of University of Chicago Library, Department of Special Collections

people were attacked, and sixty-four of these were killed and eaten or died as a result of their injuries. The records also show that the majority of attacks were on children. In September, 1765, one animal was killed; the second was killed in June, 1767. The attacks stopped.

Both animals were exceptionally large. The first weighed 109 pounds (50 kg) and the second 130 pounds (59 kg), but because they were weighed between eight and ten hours after they were killed, the bodies must have dehydrated. Had the animals been weighed immediately after death, the smaller animal would probably have scaled 119 pounds (54 kg), and the larger about 145 pounds (66 kg). In any event, even the dehydrated weights, coupled with the measurement of the skull of the bigger animal, established that they were both much larger than the average European wolf. The attacks, and the size of the animals, suggest that they were probably mastiff-wolf hybrids, an opinion that is strengthened by descriptions of their coats, which were a mixture of chocolate brown and rufus fur, hues that are not normally found in wolves' coats.

Earlier, during the Middle Ages, European forests had been dramatically reduced, causing wolves to seek more and more domestic stock. At the same time, wandering dogs, some of which had become feral, were themselves preying on farm animals, as were large populations of mostly feral domestic cats that inhabited barns and garbage dumps. Major increases in the numbers of red foxes, badgers, and polecats all combined to cause a dramatic spread of rabies, a disease that was not at the time recognized and was fatal to all who contracted it. It seems likely that, apart from attacks by hybrid wolf-dogs and feral dogs, rabies was probably responsible for a number of attacks by true wolves. And, of course, because the survivors of such attacks later died a dreadful and agonizing death, the fear and hatred of wolves must have increased dramatically.

In the past eight years, since we adopted Tundra and Taiga, we have had over 3,000 visitors. Of that number, three were undeniably afraid of the wolves, despite the fact that the animals remained in their enclosure and could not possibly have attacked even if they had wanted to. However, six of the visitors were suffering from severe stress. In every one of those cases, Tundra would almost certainly have attacked if he had been given an opportunity to do so.

The problem with stressed individuals, or individuals who are on steroid medication, is that the wolves recognize their behavior as being abnormal. They discharge large quantities of pheromones, particularly adrenalin, and their movements, the way they actually carry themselves, and the way they walk tell the wolves that they are either afraid or ill. Yet they continue to advance because they feel reassured by the fence wire.

In wolf terms this is a contradiction. Body language is important in their world. By watching the way an animal moves — be it a

A wild wolf from southern Europe.
© William Pator/NHPA

prey species or a predator such as a cougar, bear, or lynx — as well as by scenting the pheromones that the animal is releasing, wolves can determine whether it is aggressive or fearful. If aggressive, wolves will treat it with caution; if fearful, the animal will try to run away and the wolves will attack.

Of course, wolves in the wild will nearly always avoid humans. For unknown reasons, wolves do not see humans as prey, at least in North America; and I would venture that outside of rabies or cross-breeding with large dogs, Eurasian, European, and Asian wolves are also most unlikely to attack humans.

WOLVES AS PETS

Many people ask me if wolves make good pets. I invariably answer "No!" and quickly add that no wild animal makes a good pet, especially when well-meaning people who are ignorant of the behavior of wolves or other animals adopt a young wild being and treat it as though it were a cat or a dog.

In recent years in the United States and Canada, it has become fashionable to own a wolf or a wolf-dog hybrid. According to the Humane Society of the United States, there may now be about 200,000 hybrid wolf-dogs in America, while in Canada, where no figures are available, it is not uncommon to read advertisements in Canadian newspapers offering hybrids for sale.

Supporting the postulate that hybrid wolf-dogs were mostly responsible for attacks against humans in Europe, there have been a number of such attacks in the United States. They have led to six deaths during the three years ending in the winter of 1991. There

"A long, long time ago there was a little Indian boy who liked to wander in the forest. While wandering, he came upon six wolf cubs. He looked around for the mother of these cubs, but she was nowhere to be found. So every day he would go into the forest and feed the six cubs. He often would day-dream that one day he would be a great hunter, like his father. Finally, he thought to himself, "I will take the name of the wolf, which is Sta-Ka-Ya." As the years rolled by he grew up to be a husky brave . . . and became a legend for his great exploits as a hunter."

YVONNE SAM, TOLD BY HER FATHER, IN *TALES FROM THE LONGHOUSE* BY INDIAN CHILDREN OF BRITISH COLUMBIA

A wolf-dog hybrid. Many people in North America are buying such crosses as "exotic" pets.
© Mike Biggs

have been many additional attacks on people by hybrids, many of them serious and inflicted on children.

The Humane Society of the United States is totally opposed to the unbridled practice of breeding wolves with dogs. In a 1991 news release, the agency said: "Wolf hybrids are unsuitable as pets and should be reclassified as exotic animals under state law."

"People have this romantic notion of getting back in touch with the wild, but what they end up with is a creature that's a dangerous psychological mess because it doesn't know whether to be a wolf or a dog," said Sandy Rowlands, the society's Great Lakes regional director.

Dr. Randall Lockwood, the society's vice-president for field services, noted: "Wolf hybrids undo 15,000 years of domestication that made the wolf into the dog, capable of being a safe and happy companion in our homes. Keeping hybrids does nothing to improve dogs or the image of the wolf."

The main problem with buying a hybrid is that the prospective owner does not really know the animal's genetic mix, although it is likely that the breeder, anxious to make a sale, will assure a prospective buyer that the pup is 90 percent wolf, or, conversely, if the prospective buyer is concerned about having too much wolf in the pup, the breeder will likely assure the buyer it is only 25 percent wolf. In any event, the result of such breeding produces an animal that, as the saying goes, "is neither fish, flesh, fowl, nor good red herring."

Another problem is that the majority of hybrid wolf buyers know little, if anything, about the animal's needs, its future behavior, and its response to human discipline. Wolves, for instance, *must* be disciplined in wolf fashion. This means that their keeper must understand wolf hierarchy and must, as best he or she can, deal with the animal on its own terms. It can be done, of course, but those who have not really studied the behavior of wolves stand little chance of doing it. The result is usually "bad news" at some point or another.

No one should ever consider tying up a wolf or a wolf hybrid. Nor should such animals be kept in a small enclosure, especially one that has no treed shelter and offers only a cement floor. Animals kept in such conditions, whether wolves or wolf-dog crosses, will respond badly to their treatment. They may become inordinately fearful, or they may become extremely aggressive.

In addition, these animals should never be allowed to play with small children. There have been many tragedies in that situation. In March, 1990, in Anchorage, Alaska, a four-year-old girl was attacked by a hybrid while her parents were nearby. She was grasped by the head and shaken violently. Much of her scalp was ripped away and her face was bitten. Also in the spring of 1990, again in Alaska, a hybrid that was credited with loving children attacked a four-year-old boy, broke his arm, and mauled his chest and face. As well as attacks on humans, feral hybrids in different parts of the United States and Canada have been responsible for killing livestock, thus reviving the centuries-old conflict between man and wolf.

The wolf-dog hybridization problem is further exacerbated by the mating of wild dogs and wolf-dog crosses with coyotes. Occasionally, lone wolves that have been deprived of their packs by hunting and extermination policies, and are therefore unable to breed with members of their own species, also mate with feral dogs or with dog-coyote hybrids.

For these reasons, it is probable that through human ignorance

A northern husky dog nurses her pups. Of all the canids, huskies and malamutes are probably the most like wolves in looks and habits. They may be the closest wolf relatives.
Fred Bruemmer

and mismanagement, a new breed of predator is likely to emerge on the North American continent. This animal will most likely survive in rural-urban areas, will prey heavily on domestic stock, and, because of its genetic mix, will most likely be vulnerable to rabies, thereby spreading the disease more rapidly than it is already spreading outside of the wilderness.

Perhaps the present wolf-dog fad stems in part from the knowledge that several thousand years ago the aboriginal people of northern North America and northern Eurasia evidently domesticated the wolf. They used it as a pack animal, to help in hunting and to pull a sled. In time, the huskies of northern Canada and Siberia and the malamutes of Alaska came into being. Both breeds can, and do at times, continue to mate with wolves. But people in the southlands should not think that these northern hybrids will make good pets. They do not. They are working animals — or they were before the advent of motorized snow sleds, which have now largely replaced dog transportation in the north. In any event, most of the northern huskies or malamutes are at least half wild and have, in fact, been known to attack and kill their owners. About ten years ago, a woman in the Northwest Territories was training her team of dogs to run in the annual Alaskan sled-dog races. Breaking the trail on snowshoes for her team, the woman fell. The dogs piled on top of her and killed her.

The northern wolf-dogs should not, of course, be confused with the pedigreed huskies and malamutes that are popular outside of the northlands as pets, as show dogs, and as racing dogs. None of those animals have recent wolf genes.

Is it not ironic that the wolf, an animal that has been hated for so long, has now become a coveted "pet," albeit that it must first be altered with the genes of the domestic dog? It is also sad that some people, especially status seekers and those who want to present a *macho* image, feel that the pup for which they may have paid as much as a thousand dollars should be bullied into submission, so the keeper may become the animal's master.

Wolves are not averse to water. They readily cross streams, as this one is doing, and in the summer they often bathe in shallows to keep cool.
© Erwin and Peggy Bauer

Bullying may turn those hybrids with a low percentage of wolf genes, or some pure wolves that in a family pack would be destined to become underlings, into servile, neurotic wrecks. If such behavior is directed towards other animals, trouble lies ahead. When a person buys a hybrid that is mostly wolf and seeks to dominate it with brutal methods, the hybrid may turn vicious, if not towards its tormentor, then towards visitors.

In October, 1991, when we agreed to accept Silva, the four-month-old wolf that had been the captive of a motorcycle gang, she should, at that age have weighed between twenty-five and thirty pounds (11 and 14 kg). When she arrived, she weighed twelve pounds (5.5 kg). She had been tied by the neck to a heavy log, which she had to tow around a small, cement-floored enclosure. As a result, her thigh and buttocks muscles were not developed: she hopped, kicking with both back legs simultaneously. She also had rickets, which had resulted in a calcium mass on the inside of her right thigh bone, and her coat was a mess. She had no guard hairs, and her suit of underfur looked more like that of a fleeced sheep.

Beyond such physical problems, Silva was highly stressed, to the point where she had become an elective mute, perhaps because when she had cried as a pup, she had been beaten into silence. She presented, in fact, all of the stress symptoms seen in children that have been highly abused. When I described her behavior to Dr. Alyn

Roberts, a clinical psychologist friend of ours who practices in Madison, Wisconsin, he fully agreed that she presented the classic symptoms of human juvenile stress.

When she first arrived, we had to house Silva in one side of our basement, which had an earthen floor that had never been paved. A wall separated that part of our understory from the main section, and a small doorway gave access to it. Here, with Sharon lavishing attention on her, the little wolf began to respond, although she remained shy of me throughout her time there, except when I lay flat on the earth, on my back, and did not move; then she would approach cautiously, sniff me, and perhaps nibble at my shoes, or pull on a trouser leg.

Later, working alone, Sharon carried the wolf bodily from the basement, crawling through a small window to do so, and placed her in a small enclosure attached to the north side of our home, an area twenty by thirty feet (6 by 9 m). It was grassed and had a comfortable shelter. For a wolf, 600 square feet (54 m^2) of enclosure is not enough, but we had to have her near us at all times, and we had to be able to give her the medication that she desperately required.

By this time, Silva tolerated me and even followed me around her enclosure on occasion, nipping playfully at my pant legs. When I was outside the enclosure, she would solicit play — bowing, wagging her tail, and then dashing away, a wolfish invitation for a chase. But if I entered the enclosure, she became frightened and then she dashed round and round the fencing, her expression one of fear. Again, I would lie prone, even in midwinter, and let her approach me, investigate me with her nose, perhaps lick my fingers. In such ways, she would occasionally allow me to stroke her chest. We were making progress, but if male visitors arrived, she would immediately race around her enclosure. For that reason, we did not allow men near her pen.

With women, Silva was comfortable. When women visited the enclosure for the first time, the little wolf would trot right up to them, lick a proffered hand, and then solicit play. With Sharon, of course, she was inseparable. She played with her, climbed on top of her, stole her gloves, chewed up her shoelaces, and generally loved her.

By January, 1992, the rickets growth had disappeared from her thigh. She had put on weight and had started growing guard hairs; they were rather sparse, but it was a start. On the other hand, she had not yet uttered a single sound. Was she, indeed, an elective mute, or had her vocal chords been damaged by the neck rope she had been forced to use to haul her imprisoning log? From where she was, she could see and hear Tundra and Taiga, both of whom often howled, whined at each other, howl-barked when they felt like it, and growled in mock-combat. Tundra and Taiga could also smell Silva, and she could smell them, for she had given us many signs that her olfactory system, as well as her hearing, were as acute as those of the two other wolves.

One morning in spring, while Tundra and Taiga were howling, Silva began to respond. She sat on her haunches, facing the big enclosure, head up, mouth opening and shutting as she replied. It was the classic wolf pose, except that what came out of her throat was not a full-blooded howl, but a bird-like squeak. As time passed, her voice improved. At this writing, she has not yet barked, although she can howl (after a fashion) and she has learned to whine with pleasure.

Before she arrived, we had hoped that we could introduce Silva

Although snoozing, this wolf's ears are pricked upright, on alert.
© Wm. Munoz

to Taiga and Tundra, for wolves, especially females, usually dote on pups, even strangers. But when she arrived, we realized immediately that we could not possibly put a collar and lead on her in order to guide her to the large enclosure where the business of introducing her to the older wolves would take place. Nevertheless, we continued to hope, thinking that perhaps after a month or two with us, we would be able to carry out our original intention. It was not to be. Although Silva put on weight, became close to Sharon, and even invited me to play with her now and then, she still would not allow herself to be grasped in any way. So, she stayed where she was while I pondered the construction of her ultimate quarters. She could not remain in the small enclosure much beyond April or May. She needed space.

Then, through the auspices of the Ontario Society for the Prevention of Cruelty to Animals and Audrey Tournay, a soft-hearted expatriate English lady who runs a wildlife sanctuary not too far from where we live, we agreed to accept an arctic wolf female that had also been brutalized. By this time, we had erected a large 12,000-square-foot (1100 m^2) enclosure in the more forested part of our property, and I decided that the newcomer would go there. The enclosure was divided in two by a fence, with a sliding gate that would permit communication between the two sides. We had hoped that Silva would share it with another animal. I had not considered placing her there on her own because I was afraid that isolated in such a milieu she might again become seriously stressed.

In any event, the newcomer arrived May 16, 1992. What a pathetic sight! Her tail had been entirely bitten off and her right ear half-bitten off during a fight with six large German shepherd dogs

"The old leader of the wolves, he come out, left the pack and he come out under the tree, and he seen me up there. He howled a few times, and then they give that lonely wolf call, and the rest of them all come back again. And he led them away on to a big meadow, and they dug out a beaver house . . . Well, he got them to work and opened it up, and do you know what they were doing? They were gettin' the beaver to come and chew the tree down to get me. Yeah! I just sat there and kept quiet, and they came and they chewed and chewed at the tree till down went the tree. But I was gone. And I made it back to my camp."

JOE THIBADEAU QUOTED IN *FOLKLORE OF CANADA* BY EDITH FOWKE

(or so we were told). Her left thigh had been broken and although it was now healed, it had not been set properly, with the result that she walked stiffly after a nap. In addition, after I had sent a sample of her stool to our capable veterinarian, Dr. Laurie Brown, we discovered that she had whipworm and hookworm. We medicated her immediately to rid her of those parasites.

Apart from her injuries and parasites, the wolf was in dreadful physical condition. Her coat, like Silva's when she first came to us, looked moth-eaten and sheep-like. No guard hairs had grown. Hair needs protein in order to grow and a starved animal cannot spare its protein for such a luxury. She had been fed on cheap, dry dog food, which should never be given to carnivores, because it goes right through them without conferring much nutritional benefit. If all that were not enough, Alba, as I decided to call her because of her color (*albus* means "white" in Latin), had evidently been savaged by the dogs the day before she came to us. She had a number of superficial cuts around the muzzle and a bad bite between two toes of her right front foot, which was swollen, red, and obviously painful; but the day after we gave her three antibiotic tablets, the wound was almost healed, testifying once again to the stamina of wolves, and their capacity to heal quickly.

We fed Alba a diet of raw meat three times a day. Indeed, on her first day with us, she ingested ten pounds (4.5 kg) of meat, a mixture that consisted of beef, chicken, beef liver, minced-meat balls, and marrow bones, as well as multivitamins, calcium, and cod liver oil. I don't know how much she weighed when she reached us, but by lifting her bodily, I estimated that she probably weighed between eighty and eighty-five pounds (36 and 38 kg).

Arctic wolves are somewhat shorter in the body than their southern relatives, and they have dark hazel rather than yellow eyes.
David Mech

Alba's personality was — and continues to be — absolutely astonishing. Despite the savage treatment that she had received and despite her poor diet and her parasites, she is loving and gentle. Although hardly physically "pretty" because of her past treatment, she is a temperamentally beautiful animal. How she turned out to be that way is beyond my understanding. She is without a doubt the gentlest wolf I have ever had anything to do with.

Wolves are inherently affectionate. Tundra and Taiga, for example, who see me as the leader of their pack, and Sharon as a senior, benevolent "aunt," always demand affection and praise whenever we enter their enclosure, which is usually at least twice each day. Like all of their kind, they need family reassurance. But they are rarely gentle, even when licking our hands or leaning on one of us so we can pat them and scratch behind their ears. In leaning, they almost knock one down, not a difficult task for them when it is considered that Tundra weighs approximately 120 pounds (55 kg) and Taiga is only about ten pounds (5 kg) lighter. Quite often, when my back has been turned, Tundra has reared on his hind legs and thumped down on my shoulders with his huge paws before thrusting an inquisitive nose into my neck and slapping his long, wide tongue on my face.

Alba, however, is always gentle. She approaches quietly, leans against us and stands still while Sharon brushes her or while I stroke her. She rarely licks, but she likes to reach for Sharon's hand or arm and hold it in her mouth, which is something that she does not do to me, probably because she sees me as the alpha male.

Two weeks after Alba's arrival, we decided that it was time to move Silva to the other half of the enclosure. We were not looking

forward to it. Despite her close relationship with Sharon, the young wolf did not allow herself to be held for more than a split second, and certainly did not allow herself to be picked up nor to have a collar fastened to her neck so that she could be led. We probably could have forced the issue, but we were afraid that, if we did, she would become stressed all over again.

Once again we called on Dr. Brown, who arrived accompanied by animal health technician Debbie Barry. We decided that Sharon would give Silva an injection containing a small dose of anesthetic. The injection was intended to tranquilize the wolf and allow Dr. Brown to anesthetize her completely, not only so that she could be moved without causing stress, but also so that Dr. Brown could examine her. Things did not quite work out that way.

First, it took Sharon about half an hour to inject a minute amount of drug into the wolf; then, when it seemed that Silva was, indeed, at least partially tranquilized, Dr. Brown took a turn. As soon as she was approached, Silva recovered, and began to stagger around the enclosure. Dr. Brown tried again, and managed to inject a little more drug into her. It was still not enough. Although she lay as though she were fully tranquilized, when approached she immediately got up, her adrenalin pumping to dampen the effects of the drug, and the whole performance began anew. At last, however, Sharon tried again and succeeded. Silva was tranquilized sufficiently to allow Debbie to pick her up. Debbie carried the body, while Dr. Brown supported the head. I stayed out of everybody's way as the procession walked the hundred yards (90 m) to the enclosure, where Silva was set down in a shady spot.

The entire episode had taken over two hours and, because of

"One day in early spring, Nanabozho went on a hunt with Old Man Timber Wolf and his two sons. When evening came, they looked for a place to sleep.... In the night Nanabozho got cold. Old Man Wolf told his sons to put a blanket over Nanabozho. They put their tails over him, but Nanabozho did not like the tails. He got too warm. Then the wolves took their tails off Nanabozho. Then he was cold again. This went on and on through the night."

FROM "NANABOZHO AND THE WOLVES" IN *OJIBWA MYTHS AND LEGENDS* BY SISTER BERNARD COLEMAN ET AL

Wolves feeding at a kill occasionally engage in minor arguments.
© Fred Harrington

the animal's resistance to the anesthetic, Dr. Brown was not able to give her the intended physical examination. Inasmuch as Silva is an active, young wolf who is playful and full of mischief, we will have to assume she is healthy.

Looking back on the history of the wolf, there are few mammals that have had a more powerful influence on humans. Even at the height of their persecution, however, wolves were revered in many parts of the world, and although a considerable number of humans sought to kill as many of them as they could, there were others who admired the animal and even took on its name and dressed in wolfish garb on ceremonial occasions.

Chapter Seven

WORLD STATUS OF WOLVES

THE WOLF HAS THE GREATEST NATURAL RANGE OF ANY LIVING LAND mammal other than man. At one time, wolves were found throughout the northern hemisphere, especially in Eurasia and North America, where they thrived from as far north as Canada's Ellesmere Island, only 400 miles (600 km) from the North Pole to as far south as the mountain regions of Mexico. The only barriers to the wolf's expansion were the desert lands and the torrid jungle country in Mexico's south. Wolves continue to exist in twenty-seven countries, although in most of these they are perilously close to extinction.

Reporting on the wolves of Europe (*Canis lupus lupus*) to the International Wolf Symposium that was held in Washington, D.C., in May, 1987, Luigi Boitani of the Department of Biology, University of Rome, noted that the wolf habitat in Eurasia has shrunk to a very small portion of its original range. He told the assembly, "...the most dramatic changes are taking place in the European and southern sections." There, the animals are forced into co-existence with humans and, of course, with the environmental exploitation and ecological change created by the increase in human numbers.

The wolf in Italy, Dr. Boitani believes, is one of the best examples of wolf-human contact. And the Italian wolf has been one

After vigorous activity or nursing pups, wolves require large quantities of water. They lap up water using a scooping motion of the tongue.
© Thomas Kitchin

of the most studied in Eurasia during a continuing series of biological investigations that have been monitoring it since 1972. Supported by the World Wildlife Fund, the Italian study began with a wolf census and continued monitoring by means of radio telemetry after a number of wolves had been live-trapped and fitted with a radio collar. By 1973, it had become evident that no more than a few hundred wolves continued to survive in Italy, and their range had been restricted to the high mountain area of the central and southern regions of the country.

An Italian wolf from Abruzzi National Park. Only about 250 wolves survive in Italy.
© Michael Leach/NHPA

Because the wolf's original prey had been brought to extinction in Italy by the end of the nineteenth century, the surviving wolves today feed mostly on refuse dumps near hinterland villages, travelling singly or in pairs rather than in packs, although Dr. Boitani believes that in daylight they gather together to rest as a pack.

Occasionally, although sheep farmers guard their animals carefully, wolves will take one or two sheep. They also sometimes take cattle, mostly calves. As has occurred throughout history, when the wolf attacks human-owned livestock, there is a major outcry. This impulse to retaliate, say the researchers, poses the major threat to wolf survival, although in more recent times the Italian government, prodded by the wolf biologists, has enacted a law that provides full payment to farmers for wolf damages. The new law also forbids the use of poison baits.

As a result of active conservation efforts, Dr. Boitani reported to the symposium that wolf numbers during the past ten years have increased 50 percent. Nevertheless, the small population of Italian wolves gives the biologist cause for concern that the wolf population will remain between 200 and 300 animals.

Opposite: *Former and present distribution of the wolf.*

Source: J.R. Ginsberg and D.W. Macdonald, *Foxes, Wolves, Jackals and Dogs: An Action Plan for the Conservation of Canids,* IUCN/SSC Canid Specialist Group and IUCN/SSC Wolf Specialist Group (L.D. Mech, Chair), (Gland, Switzerland, 1990), p. 38 and 40.

An Israeli wolf. The two subspecies of wolf in the Middle East are highly threatened. © Jean-Paul Ferrero/ Auscape

Survival of the wolf in Italy is also threatened by the high number of feral and stray dogs that roam freely throughout the entire wolf range, according to Dr. Boitani. They compete for food and for territory. Interbreeding between wolves and dogs has also been confirmed. Dogs are preying on livestock, but the wolves are blamed for most of the damage. There are psychological reasons why people blame wolves, but it also occurs because in certain provinces only wolf damages are refunded.

"Wolf survival will ultimately depend on the solution of the feral and stray dog problems, an intricate issue which has deep historical, ecological and psychological aspects," Dr. Boitani concluded.

In 1992, the plight of the Italian wolf remains the same. Its status is considered lingering, with a population of 250 animals. It is highly threatened, according to the International Union for Conservation of Nature and Natural Resources (IUCN).

The wolves in Spain (*Canis lupus signatus*), although their numbers are estimated to range from 500 to 1,000, are also facing extinction in the not-too-distant future. They have a low population density and have been placed in the "threatened" category by the IUCN, which notes that they now occupy only 10 percent of their original range.

The Spanish wolf has been given partial legal protection, but little effort is made to enforce this law. Meanwhile, destruction of the wolf's habitat and continued persecution will, unless brought to a halt soon, cause the animal's extinction.

The main prey for these wolves consists of livestock, especially sheep, and of wild boar and roe deer. Shepherds kill wolves at every opportunity, engaging in chases and shooting any wolf that comes

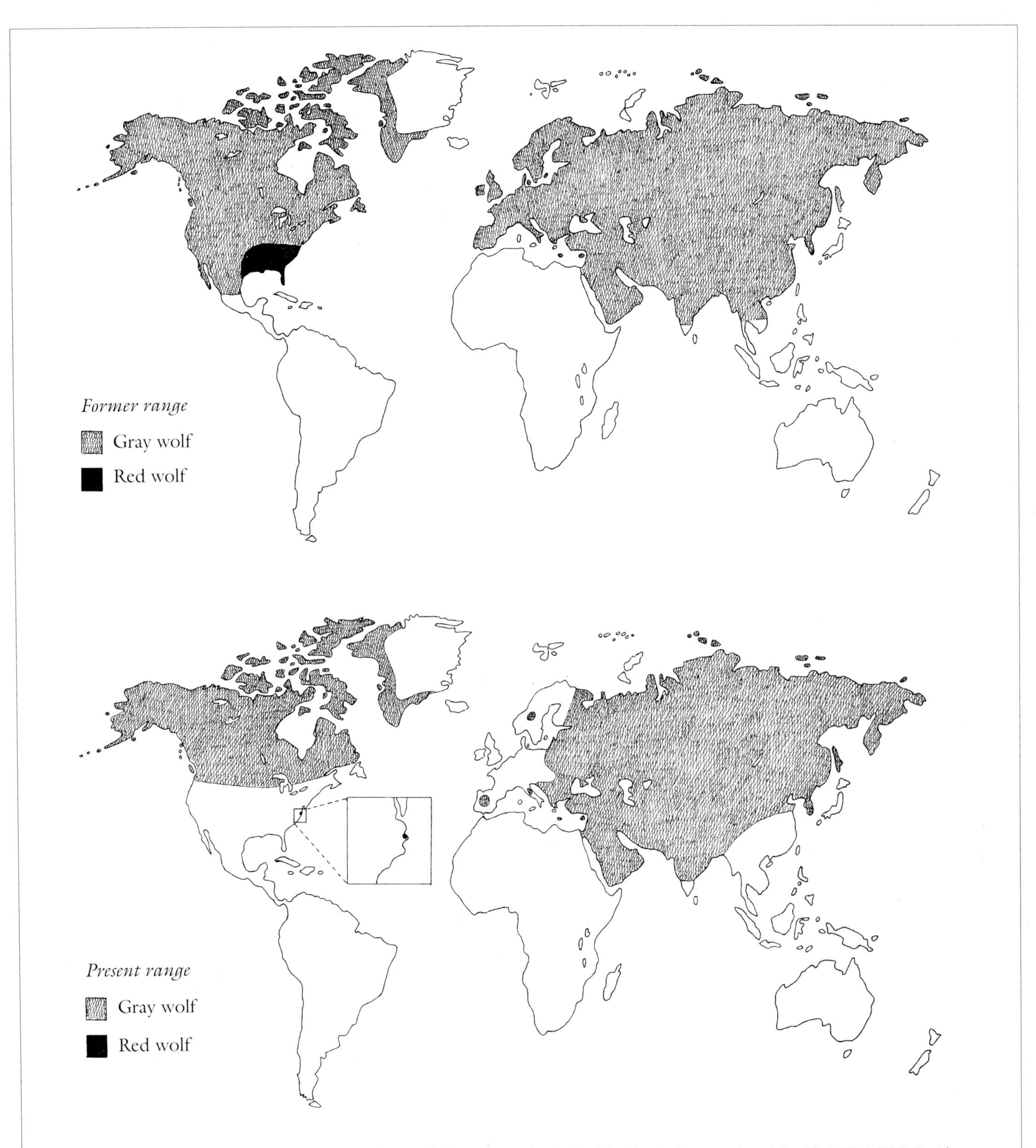

Source: J.R. Ginsberg and D.W. Macdonald, *Foxes, Wolves, Jackals and Dogs: An Action Plan for the Conservation of Canids,* IUCN/SSC Canid Specialist Group and IUCN/SSC Wolf Specialist Group (L.D. Mech, Chair), (Gland, Switzerland, 1990), p. 38 and 40.

An Israeli wolf. The two subspecies of wolf in the Middle East are highly threatened.
© Jean-Paul Ferrero/ Auscape

Survival of the wolf in Italy is also threatened by the high number of feral and stray dogs that roam freely throughout the entire wolf range, according to Dr. Boitani. They compete for food and for territory. Interbreeding between wolves and dogs has also been confirmed. Dogs are preying on livestock, but the wolves are blamed for most of the damage. There are psychological reasons why people blame wolves, but it also occurs because in certain provinces only wolf damages are refunded.

"Wolf survival will ultimately depend on the solution of the feral and stray dog problems, an intricate issue which has deep historical, ecological and psychological aspects," Dr. Boitani concluded.

In 1992, the plight of the Italian wolf remains the same. Its status is considered lingering, with a population of 250 animals. It is highly threatened, according to the International Union for Conservation of Nature and Natural Resources (IUCN).

The wolves in Spain (*Canis lupus signatus*), although their numbers are estimated to range from 500 to 1,000, are also facing extinction in the not-too-distant future. They have a low population density and have been placed in the "threatened" category by the IUCN, which notes that they now occupy only 10 percent of their original range.

The Spanish wolf has been given partial legal protection, but little effort is made to enforce this law. Meanwhile, destruction of the wolf's habitat and continued persecution will, unless brought to a halt soon, cause the animal's extinction.

The main prey for these wolves consists of livestock, especially sheep, and of wild boar and roe deer. Shepherds kill wolves at every opportunity, engaging in chases and shooting any wolf that comes

Current Status of Wolves in North America

Country/State Prov./Territory	Status	Range Occupied	Cause of Decline
Alaska	6,000, viable	100%	Some control programs active
Alberta	4,000, viable	80%	Habitat loss, persecution, agricultural conflicts
Labrador	Unknown, viable	95%	No data available
Manitoba and Saskatchewan	Unknown, viable	70%	Persecution in south, habitat loss, agricultural conflicts
Mexico	Fewer than 10, highly endangered	Less than 10%	Protection not enforced, persecution, habitat destruction
Michigan and Wisconsin	35, highly endangered, lingering	10%	Persecution, habitat destruction
Minnesota	1,200, viable	30%	Persecution, habitat destruction
Newfoundland	Extinct	Nil	No data available
Northwestern United States	30, highly endangered, slowly recolonizing	5%	Persecution, habitat destruction
Northwest Territories	5,000–15,000, viable	100%	Stable
Ontario, Quebec	10,000?, viable	80%	Persecution, habitat loss, agricultural conflicts
Southwestern United States	Extinct	Nil	Persecution, habitat destruction
Yukon/B.C.	8,000, viable	80%	Habitat loss, persecution, agricultural conflicts

Current Status of Wolves in Europe and Asia

Country	Status	Range Occupied	Cause of Decline
Afghanistan	1,000?, in decline	90%	Unknown
Albania	Unknown	Unknown	Unknown
Arabian peninsula	Fewer than 300, in decline	90%	No protection, persecution
Bhutan	Unknown	Unknown	Unknown (but protected)
Bulgaria	100?, lingering, highly threatened	Unknown	No protection, persecution, habitat destruction
China	Unknown	20%	Persecution, extermination efforts, habitat destruction
Czechoslovakia	100?, lingering, in steep decline, endangered	10%	No protection, persecution, habitat destruction
Egypt (Sinai)	30, highly endangered	90%	No protection, persecution
Europe (Central)	Extinct	Nil	Persecution, habitat destruction
Finland	Fewer than 100, lone wolves and pairs	Less than 10%	Persecution, no protection
Greece	500+, viable but declining	60%	Persecution, habitat destruction
Greenland	50?, lingering, threatened	Unknown	Persecution
Hungary	Extinct	Nil	Unknown
India	1,000–2,000, lone wolves or pairs, endangered	20%	Decreasing prey, persecution, unenforced protection
Iran	More than 1,000, viable	80%	Persecution
Iraq	Unknown	Unknown	Unknown
Israel	100–150, lingering, highly threatened	60%	Habitat destruction, persecution
Italy	250, highly threatened	10%	Persecution, habitat destruction, extermination of prey
Jordan	200?, highly threatened	90%	Persecution, no protection
Lebanon	Fewer than 10, highly endangered	Unknown	No protection, persecution
Mongolia	10,000+, viable, in decline	100%	Active extermination efforts
Nepal	Unknown	Unknown	Unknown
Pakistan	Unknown	Unknown	Unknown
Poland	900, viable	90%	Persecution, habitat destruction
Portugal	150, lingering, highly threatened	20%	Persecution, habitat destruction
Romania	2,000?, declining	20%	No protection, persecution, habitat destruction
Spain	500–1,000, threatened	10%	Persecution, habitat destruction
Sweden/Norway	Fewer than 10, highly endangered	Less than 10%	Persecution
Syria	200–500, lingering, low density, highly threatened	10%	No protection, persecution
Turkey	Unknown, viable, in decline	Unknown	No protection, livestock predation
former USSR (Asia)	50,000, viable	75%	Control programs everywhere, persecution, habitat destructio
former USSR (Europe)	20,000, viable	60%	Control programs everywhere, persecution, habitat destruction
Yugoslavia	2,000, in steep decline	55%	Persecution, habitat destruction

Source: J.R. Ginsberg and D.W. Macdonald, *Foxes, Wolves, Jackals and Dogs: An Action Plan for the Conservation of Canids,* IUCN/SSC Canid Specialist Group and IUCN/SSC Wolf Specialist Group (L.D. Mech, Chair), (Gland, Switzerland, 1990), p. 38-39.

into rifle range. The chase at times ends when a mother wolf has been killed and the pups are removed from their den, after which they will be taken as exhibits from village to village to show the daring efficiency of the hunters as well as to let the villages know that they have revenged the loss of the sheep. Exhibition of the cubs continues until the little animals die of starvation and dehydration, for they are not fed or watered during the tour.

This coyote is baffled by the armor of an armadillo.
© Mike Biggs

The situation in Portugal is even worse than in Spain and Italy. There are perhaps 150 individual wolves (*Canis lupus signatus*) in the country and although an international conservation association, Grupo Lubo, is fighting for them and has managed to fund a sanctuary, the animals are intensely hated and killed whenever an opportunity presents itself. Like the Spanish wolf, the Portuguese wolf has been given partial protection, but here again, there is little or no enforcement.

It is ironic to note that in France, in the region where the Beast of Gévaudan and its companion killed about a hundred people, there is a pack of captive wolves in the parish of Gévaudan. The animals are kept in a large enclosure and visited by many tourists.

More recently, Brigitte Bardot, the film-star conservationist, imported into France eighty Mongolian wolves (*Canis lupus chanco*) through her Bardot Foundation. The wolves, originally one hundred of them, had been bought by Hungarian interests and held in Budapest while awaiting transportation to an abattoir, where they were to be killed, skinned, and their pelts sold to the fur industry. The mayor of Budapest, however, was outraged by the event. He contacted the Bardot Foundation and noted: "Ces loups sont vous" ("These wolves are for you"). In due course, three trucks loaded with

The Indian wolf is considered endangered.
© E. Hamumantha Rao/NHPA

wolves set out from Budapest on a five-day journey to France, where they were received by Bardot and housed on a twelve-acre (5-ha) refuge at the Lozere animal park in Sainte-Lucie. Apart from captive animals, the wolf in France is extinct, although almost every autumn there are reports of sightings in the northern regions of the country.

About ten wolves continue to survive in Norway, and although they have been afforded full protection status, this is not being enforced by the government and survival of the animals is doubtful. Yet there are some people who continue to press for their protection, including philosopher Arne Naess and biologist Ivar Mysterud, both of the University of Oslo, who in 1987 wrote a joint paper titled *Philosophy of Wolf Policies I: General Principles and Preliminary Exploration of Selected Norms.* In their abstract, Naess and Mysterud note:

> We, as philosopher and biologist, here present some preliminary explorations of values and norms of importance in the wolf-man relationship. The presentation is centered around problems as we see them from the modern wolf range in Norway, where there should be a mixed community of sheep, wolves and men. At present we have 3.2 million sheep, 4.1 million men [people] and five to ten wolves. The wolves are confined to a small area containing small scattered sheep farms. The owners, with local approval, do not accept the wolves. . .

In their paper, the writers plead for wolf protection, noting: "Without the slightest doubt, we recommend wolves as members of the nordic life community." The paper concludes with both misgivings and a promise to continue exploring and writing about the subject.

Coyotes feed largely on smaller mammals, although they do hunt deer.
© Erwin and Peggy Bauer

Until the unification of Germany, only a very few wild wolves were reported to live in the western part of the nation; five were killed on different occasions, and sporadic, unconfirmed reports of sightings continued to surface. In addition, a few captive wolves escaped from a study area some ten years ago and they were also killed. Since unification, however, there have been some reports of sightings in the extreme northeast, in an area near the Polish border that lies between Pasewalk in the north and Eberswalde-Finow in the south and encompasses some 450 square miles (1200 km^2). None of these sightings have been officially confirmed. Apart from escaped wolves, all other sightings were of wild wolves that had crossed the border from Poland.

In North America, it has been estimated that there are between 35,000 and 40,000 wolves, the majority of which are in Canada. In the United States, Alaska has the highest population, with about 6,000 wolves.

Soon after the turn of the twentieth century, wolves were almost entirely extirpated from the forty-eight states of the United States and were greatly thinned in southern Canada. The coyote, a wily, opportunistic little canid, took full advantage of the situation.

Unlike wolves, coyotes do not control their numbers. When a pair mates, a female in good condition may give birth to as many as eighteen pups, although because she has only eight teats, it is highly unlikely that all of them will survive. Nevertheless, on average, five to six pups are born to a pair each year, and those that survive into their second year will likely mate and have their own young. Unlike the wolf, which lives in a pack in which only the leaders usually mate, coyotes tend to multiply geometrically — that is, 2, 4, 8, 16, 32, and so on.

These wolf skins hanging on a line in northern Canada are for sale. They are reminders of the war against wolves that humans have been waging for centuries.
© W. Perry Conway

Today, the coyote's range has expanded dramatically in many areas of Canada and the United States. Coyotes used to live in relatively open country from Central America to the prairie regions of the United States and Canada and were not known in Ontario and in other forested regions of North America until about 1908. Since then, however, they have moved into ranges where wolves have been thinned out through persecution. Now they have invaded most Canadian provinces as well as the northern states and the eastern seaboard of the United States. They have also reached Alaska and the southern regions of the Yukon Territory.

The situation in Eurasia and in the other parts of the world where wolves are still found is quite different from that in North America. There are no coyotes in those regions, although in some parts of Scandinavia there are individuals who are breeding the raccoon-like dog, *Nyctereutes procyonides*, from Siberia and Asia. This animal has a coat that is used by the fur industry, and it breeds well. It could become a problem if it is allowed to escape and to crossbreed with domestic dogs.

Control programs are still being employed in Alaska, especially shooting from aircraft and trapping. In Canada, control programs and trapping continue in all provinces and in the Yukon and the Northwest Territories. In addition, wolves are legally trapped for their fur and hunted as trophies, and they are also poached for sale to trophy buyers in Canada, the United States, and Western Europe.

Despite these depressing facts, there is reason to hope. In recent times, more and more people have become concerned about the persecution of this mammal and, indeed, many now look upon the wolf as a symbol of courage, endurance, and intelligence.

The Mexican wolf is classified as highly endangered.
© C. Allan Morgan

CONSERVATION ORGANIZATIONS

Canada

Canadian Wolf Defenders
Box 3480, Stn. D
Edmonton, Alta.
T5L 4J3

Wolf Awareness International
G-2 Farms
RR #3
Ailsa Craig, Ont.
N0M 1A0

World Wildlife Fund Canada
90 Eglinton Ave. E.
Ste. 504
Toronto, Ont.
M4P 2Z7

United States

Defenders of Wildlife
1244 Nineteenth St. N.W.
Washington, DC 20036

International Wolf Center
5930 Brooklyn Blvd.
Ste. 200
Brooklyn Center, MN 55429

North American Wolf Society
P.O. Box 82950
Fairbanks, AK 99708

Preserve Arizona's Wolves
1413 East Dobbins Rd.
Phoenix, AZ 85040

Wolf Ecology Project
120 Derns Road
Kalispell, MT 59901

The Wolf Fund
P.O. Box 471
Moose, WY 83012

Wolf Haven America
3111 Offut Lake Rd.
Tenino, WA 98589

World Wildlife Fund
1250 24th St. N.W.
Washington, DC 20037

International

Wolf Haven International
7447 Boston Harbour Road NE
Olympia, WA 98506

World Conservation Union
World Conservation Centre
Avenue du Mont-blanc
CH-1196 Gland
Switzerland

World Wildlife Fund
Panda House
Weyside Park
Godalming, Surrey
England GU7 1XR

Publications

WOLF! Magazine
Janet Lidle
P.O. Box 112
Clifton Heights, PA 19018

Wolf News
c/o Dick Dekker
3819-112A St.
Edmonton, Alta.
T6J 1K4

Red Wolf Newsletter
c/o Will Waddell
Research Biologist
Point Defiance Zoo and Aquarium
5400 North Pearl St.
Tacoma, WA 98407

SELECTED BIBLIOGRAPHY

Allen, Durward L. 1979. *Wolves of Minong*. Boston: Houghton Mifflin.

_____. 1963. The costly and needless war on predators. *Audubon Magazine*. 65 (2): 82-89, 120-121.

_____. 1977. Wolf research on Isle Royale. In *North American Big Game*. 7th ed. Dumfries, VA: Boone and Crockett Club and National Rifle Association of America.

Banfield, A.W.F. 1974. *The Mammals of Canada*. Toronto: University of Toronto Press.

Fiennes, Richard. 1976. *The Order of Wolves*. London: Hamish Hamilton.

Fox, Michael W. 1971. *Behaviour of Wolves, Dogs and Related Canids*. London: Jonathan Cape.

_____. 1980. *The Soul of the Wolf*. Boston: Little, Brown.

_____. 1984. *The Whistling Hunters: Field Studies of the Asiatic Wild Dog* (Cuon alpinus). Albany, NY: State University of New York Press.

Ginsberg, J.R., and Macdonald, D.W. 1990. *Foxes, Wolves, Jackals, and Dogs. An Action Plan for the Conservation of Canids*. Gland, Switzerland: IUCN.

Griffin, Donald R. 1984. *Animal Thinking*. Cambridge: Harvard University Press.

Lawrence, R.D. 1986. *In Praise of Wolves*. New York: Henry Holt.

_____. 1980. *Secret Go the Wolves*. New York: Henry Holt.

Leydet, Francois. 1988. *The Coyote: Defiant Songdog of the West*. Norman, OK: University of Oklahoma Press.

Mech, L. David. 1981. *The Wolf: The Ecology and Behavior of an Endangered Species*. Minneapolis: University of Minnesota Press.

Murie, Adolph. 1944. *The Wolves of Mount McKinley*. Washington, D.C.: U.S. Printing Office.

Naess, Arne, and Mysterud, Ivar. 1987. Philosophy of wolf policies I: general principles and preliminary exploration of selected norms. *Conservation Biology*, May, 1987.

Nowak, Ronald M. 1991. *Walker's Mammals of the World*. Vol. 2. Baltimore, MD: The Johns Hopkins University Press.

Peek, J.M., et al. 1991. *Restoration of Wolves in North America*. Wildlife Society Technical Review 91-1. Bethesda, MD: The Wildlife Society.

Peterson, Rolf Olin. 1977. *Wolf Ecology and Prey Relationships on Isle Royale*. National Park Service Scientific Monograph Series. No. 11.

Pollard, John. 1964. *Wolves and Werewolves*. London: Robert Hale.

Rutter, Russell J., and Pimlott, Douglas H. 1968. *The World of the Wolf*. New York: J.B. Lippincott.

Schenkel, R. 1947. Behaviour of Wolves. *Behaviour* 1 (2): 81-129.

Standen, V., and Foley, R.A., eds. 1989. *Comparative Socioecology: The Behavioral Ecology of Humans and Other Mammals*. Boston: Blackwell Scientific Publications.

Wayne, R.K., and Jenks, S.M. 1991. Mitochondrial DNA analysis implying extensive hybridization of the endangered red wolf *Canis lupus*. *Nature* 351: 565-568.

Young, Stanley P., and Goldman, Edward A. 1944. *The Wolves of North America*. New York: Dover Publications.

Zimen, Erik. 1978. *The Wolf: A Species in Danger*. New York: Delacorte Press.

INDEX

Numbers in italics refer to photographs or illustrations.